194

Advances in Anatomy Embryology and Cell Biology

Editors

F. F. Beck, Melbourne · F. Clascá, Madrid
M. Frotscher, Freiburg · D. E. Haines, Jackson
H.-W. Korf, Frankfurt · E. Marani, Enschede
R. Putz, München · Y. Sano, Kyoto
T. H. Schiebler, Würzburg

L. Santamaría · I. Ingelmo · L. Alonso
J. M. Pozuelo · R. Rodriguez

Neuroendocrine Cells and Peptidergic Innervation in Human and Rat Prostate

With 31 Figures and 1 Table

 Springer

Luis Santamaría, Prof. Dr.
Lucia Alonso, Dr.
Department of Anatomy, Histology and Neuroscience
School of Medicine
Autonomous University of Madrid
C/Arzobispo Morbillo, 2
28029 Madrid
Spain
e-mail: luis.santamaria@uam.es

Ildefonso Ingelmo, Dr.
Department of Anaesthesiology
Hospital Ramón y Cajal
Madrid
Spain

José Manuel Pozuelo, Dr.
Rosario Rodriguez, Dr.
Department of Medical Sciences, Histology and Anatomy
School of Medicine
San Pablo-CEU University
Madrid
Spain

ISSN 0301-5556
ISBN 978-3-540-69815-9 Springer Berlin Heidelberg New York

This work is subject to copyright. All rights reserved, whether the whole or part of the material is concerned, specifically the rights of translation, reprinting, reuse of illustrations, recitation, broadcasting, reproduction on microfilm or in any other way, and storage in data banks. Duplication of this publication or parts thereof is permitted only under the provisions of the German Copyright Law of September, 9, 1965, in its current version, and permission for use must always be obtained from Springer-Verlag. Violations are liable for prosecution under the German Copyright Law.

Springer is a part of Springer Science+Business Media
springer.com
© Springer-Verlag Berlin Heidelberg 2007

The use of general descriptive names, registered names, trademarks, etc. in this publication does not imply, even in the absence of a specific statement, that such names are exempt from the relevant protective laws and regulations and therefore free for general use.
Product liability: The publisher cannot guarantee the accuracy of any information about dosage and application contained in this book. In every individual case the user must check such information by consulting the relevant literature.

Editor: Simon Rallison, Heidelberg
Desk editor: Anne Clauss, Heidelberg
Production editor: Nadja Kroke, Leipzig
Cover design: WMX Design Heidelberg
Typesetting: LE-TEX Jelonek, Schmidt & Vöckler GbR, Leipzig
Printed on acid-free paper   SPIN 11971597   27/3150/YL – 5 4 3 2 1 0

List of Contents

Acknowledgements

We are grateful to Dr. J. Codesal, Department of Anatomy, Histology, and Neuroscience, Autonomous University of Madrid, for his contribution to the human prostate bibliography, to Dr. F. Teba, Department of Surgery (Urology), Autonomous University of Madrid, for supplying the human prostate specimens, as well as to Dr. R. Martín, Department of Pathology, Hospital N. Sra de Sonsoles, Avila, and to Dr. R. Arriazu, Department of Medical Sciences (Histology and Anatomy), San Pablo-CEU University, for scientific contributions to Sects. 2 and 5, respectively and to Dr. V. Gómez, Department of Anatomy, Histology and Neurosciences, Autonomous University of Madrid for his contribution to the rat prostate bibliography.

Abbreviations

5-HT	5-Hydroxytryptamine
APUD	Amine precursor uptake and decarboxylation
BPH	Benign prostate hyperplasia
CGR	Chromogranin
CGRP	Calcitonin gene-related peptide
CZ	Central zone
DOPA	Dihydroxyphenylalanine
EGF	Epidermal growth factor
HPLC	High-performance liquid chromatography
NANC	Nonadrenergic, noncholinergic
NEC	Neuroendocrine cell
NPY	Tyrosine neuropeptide
PAP	Prostate acid phosphatase
PGP 9.5	Protein gene product 9.5
PSA	Prostate-specific antigen
PTHrP	Parathyroid hormone-related peptide
PZ	Peripheral zone
SER	Serotonin
SP	Substance P
TH	Tyrosine hydroxylase
TRH	Thyrotropin-releasing hormone
TSH	Thyrotropin
TZ	Transition zone
VAChT	Vesicular acetylcholine transferase
VIP	Vasoactive intestinal peptide

1
Introduction

The prostate causes a significant number of medical problems in the adult male, and the lower urinary tract symptoms are accepted as an unavoidable consequence of male aging. Most of these symptoms are mainly due to clinical benign prostatic hyperplasia (BPH), which is the most frequent benign tumor in the male, independent of race or culture. On the other hand, cancer of the prostate shows an increasing incidence, being the second leading cause of death in men, after lung cancer. It has an etiology related to multiple factors: age, race, androgen dependence, chemical agents, diet, etc.

Both pathologies are very costly in terms of medical resources, and they significantly diminish quality of life. More than 400,000 prostate resections per year are done in the US, and these result in an approximate cost of 5 billion dollars per year. Because of all these circumstances, a better knowledge of the mechanisms regulating normal, hyperplastic, and neoplastic prostate growth is important for treatment and prevention of BPH and prostate cancer.

Since the last review about prostate neuroendocrine cells published in 1998 there have been new and exciting developments relating to these cells in both normal and pathologic prostate. The cross talk of signals between epithelial and neuroendocrine cells seems relevant to the development and physiopathology of the prostate; thus the relationship between these cell populations should be more deeply studied.

The prostate hosts an important number of neuroendocrine cells, whose origin and functional role need to be better addressed. These cells synthesize and deliver a number of neurosecretory substances (serotonin, neuropeptides) having regulative activities over growth, cell differentiation, and secretion. Those substances might have a remarkable influence in the development of BPH and prostate cancer.

The autonomous nervous system seems to be relevant in the maintenance of structural and functional integrity of the prostate. Thus prostatic denervation leads to extensive atrophy of the rat prostate. Besides catecholaminergic and cholinergic innervation, a wide variety of peptidergic fibers have been described in the prostate gland, such as neuropeptide Y (NPY)-, vasoactive intestinal polypeptide (VIP)-, substance P (SP)-, and calcitonin gene-related peptide (CGRP)-responsive nerves. It seems that neuropeptides may be implicated in the physiology of the prostate.

Nevertheless, the role of peptidergic innervation in prostate function is not yet well ascertained. The regulatory peptides contained in the autonomous neuroendings could intervene in prostatic secretion, but also in the development and growth of the prostate acini, modulating the action of androgens. Some neuropeptides such as VIP might be implicated in the epithelial proliferation of prostatic acini, and several investigations are relating peptidergic innervation and neuroepithelial interactions with prostatic pathologies like cancer or BPH.

The rat has frequently been employed as an experimental model to study the biology and pathology of the prostate; therefore it seems interesting to investigate the nature and distribution of peptidergic nerves during the postnatal development of rat prostate. The regulation of prostate growth has been considered as an almost exclusive function of the endocrine system; nevertheless, the finding of cholinergic and adrenergic receptors in human prostate, and the presence of abundant autonomic neuroendings, suggests a role for innervation in homeostasis, growth, and prostate function. In fact, neurotransmitter signal transduction might modulate the growth and physiology of the prostate gland, and experimental denervation causes the loss of prostate function and its atrophy.

There is evidence about morphologic and functional relationships between neuroendocrine cells and prostate nerve fibers, resulting in a neurohormonal system that might modulate androgenic action on the prostate.

Most studies try to measure the population of neuroendocrine cells or nerve fibers with semiquantitative, biased methods; therefore, a rigorous quantitation is not sufficiently addressed. Stereologic tools provide a pool of unbiased quantitative methods able to provide adequate estimations of either absolute or relative number of cells and nerve fiber sizes.

With these methods, the authors of the present work have demonstrated that the neuroendocrine cell population from the transition zone of the human prostate is greater than those in both central and peripheral regions, and this can be related to the genesis of BPH. Furthermore, we have observed that the neuroendocrine cells and the peptidergic innervation of rat prostate might be influenced by aging and androgenic status. Other factors, such as pharmacological castration or prolactin action, might also modulate the neuroendocrine-peptidergic system in the rat prostate.

2
The Human Prostate

L. Santamaría, L. Alonso

The prostate is an exocrine gland, considered the main accessory gland of the male reproductive tract (Leissner and Tisell 1979; Walsh et al. 2002). The prostate contributes 20% of the composition of semen: The prostate secretion is an opalescent fluid, rich in citrate, acid phosphatase, and proteolytic enzymes (fibrinolysins, plasminogen activator, pepsinogen II, etc.), with these secretions contributing to semen liquefaction after ejaculation and to sperm capacitation (Fawcett and Raviola 1994).

2.1
Embryology and Development

The prostate is mainly derived from the urogenital sinus, except for a portion of verumontanum, the ejaculatory ducts, and a portion of acini from the central region, which originate from Wolffian structures (Tanagho 1982). The prostate tissues appear after the seventh week of gestational age and become completely differentiated between 11th and 13th weeks. The fetal androgens are the main regulating agents for prostate prenatal growth. (Algaba 1993).

During the first years of life, the prostate gland is macroscopically unapparent, and its cells lack secretory function. Around the peripubertal stage (12 years), the gland experiences a relevant increase in the number of acini, and during puberty, the hormonal stimulus causes the sprouting of new buds of glandular ducts and there is the first evidence of secretory activity. The adult size of the prostate is reached around 20 years, with this stage being preceded by a phase of exponential growth, duplicating the glandular weight each 2.8 years (Aumuller 1991; Algaba 1993).

2.2
Gross Anatomy of the Prostate

The prostate is the biggest of all the accessory glands of the male reproductive tract. It is located in the middle of the pelvic cavity, resting on the perineal floor, behind the pubic symphysis, and surrounded by a dense connective fascia. It is cone-shaped, and slightly flattened in the antero-posterior axis, with the vertex of the cone oriented downward. The prostate from a young adult weighs around 20 g and has 25×40×30-mm mean diameters.

2.2.1
External Anatomic Relationships

- Base: In contact with the urinary bladder and seminal vesicles. The urethra penetrates into the prostate at a point near the middle of the prostatic base.
  - Apex: Located above the urogenital diaphragm and in contact with the striated sphincter of the urethra.
  - Posterior face: In contact with the rectum (prostato-peritoneal aponeurosis).
  - Anterior face: In contact with the pubo-vesical ligaments, the venous plexus of Santorini, and the pubic symphysis.
  - Lateral borders: Resting on the fascia that surrounds the elevator muscles of the anus and in contact with the vasculo-nervous tracts.

2.2.2
Internal Anatomic Relationships

- Preprostatic sphincter: It is constituted of a sheath of smooth muscle fibers, surrounding the proximal portion of the urethra. This structure might prevent the retrograde flow of the seminal fluid during ejaculation.
  - Anterior fibromuscular stroma: It is a multilayered structure of smooth muscle located on the antero-medial surface of the prostate from the bladder neck to the apex.
  - Striated sphincter: It contains striated muscle fibers, and is extended from the verumontanum to the apex.

– Prostate capsule: It is a sheath of fibrous connective tissue with smooth muscle cells deeply located and transversally oriented. The capsule sends frequent trabeculae of fibro-muscular tissue that divide the prostate parenchyma in approximately 50 lobules.

– Prostatic urethra: It is the proximal portion of the urethral duct. It presents a longitudinal folding, located in its posterior face (urethral ridge), and limited laterally by the urethral sinus; the medial portion of the urethral ridge originates the seminal colliculus (verumontanum) having a central invagination forming the prostate utriculus. In both sides of this structure are situated the outlets of the ejaculatory ducts into the urethral duct. The prostatic urethra shows an angle of 35° with the vertex in the seminal colliculus.

2.2.3
Prostate Regions

McNeal proposed a new concept of zonal anatomy of the prostate, based on macroscopic and histologic data. According to this author (McNeal 1981, 1984, 1997), the human prostate can be compartmentalized into four regions. Evidence in support of this model has been derived from morphological, histochemical, and clinical studies (Laczko et al. 2005) (Fig. 1).

– Peripheral zone: It comprises 70%–75% of the glandular tissue. The acini from this region drain to the distal portion of the urethra. Prostate cancer is highly frequent (90%) in this zone.

– Central zone: Represents 20–25% of the glandular tissue. It is partially surrounded by the peripheral zone, located in the prostate base, and is traversed by the ejaculatory ducts.

– Transition zone: Constitutes 5% of the glandular tissue in the young adult, and increases with aging. It is formed by two lobules located around the most distal portion of the proximal urethra. This region comprises also a thin layer of glands immediately surrounding the urethral duct (periurethral region). Of cases of benign prostate hyperplasia (BPH), 100% are developed in this zone. It contains few acini and abundant fibromuscular stroma. This stromal component is particularly abundant in the anterior side of the transition zone, showing very scarce acini in this location.

– Anterior fibromuscular stroma: This compartment can be considered as the most anterior component of the peripheral zone. It is constituted by a fibroconnective tissue with abundant smooth muscle fibers and without acini. The absence of an epithelial component excludes this zone from the true glandular prostate.

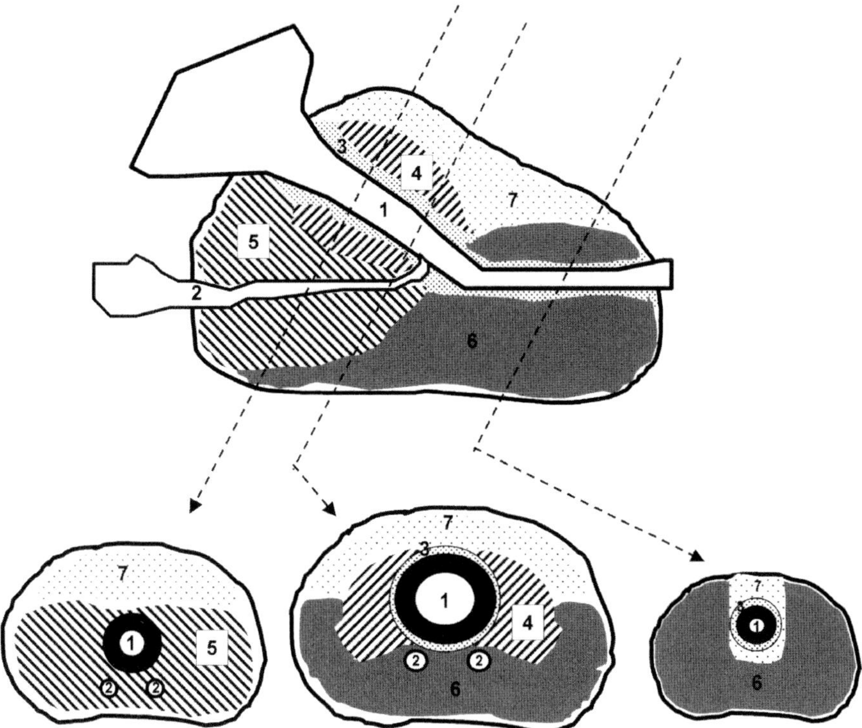

Fig. 1 Macroscopic regionalization of the human prostate, according to McNeal. The *arrows* traversing the sagittal section indicate transverse sections at three levels. (*1*) Prostatic urethra; (*2*) ejaculatory ducts; (*3*) periurethral region; (*4*) transition zone; (*5*) central zone; (*6*) peripheral zone; (*7*) anterior fibromuscular stroma

2.2.4
Blood Vessels

The prostatic arteries are originated from the pudenda interna, vesical inferior, and rectal media (hemorrhoidal). The venous circulation drains to the antero-lateral prostate plexus (Santorini), toward the seminal plexus; from there, the blood is conveyed to the hypogastric vein, via the veins from the urinary bladder. The lymphatic vessels form a periprostatic plexus on the surface of the gland that drains to the hypogastric, sacral, vesical, and external iliac lymphoid nodules.

2.2.5
Innervation

The prostate receives rich innervation from the autonomic nervous system, through the major pelvic ganglion, containing sympathetic fibers from spinal segments D_{11}–L_2; these fibers join to the major pelvic ganglion by the hypogastric nerve. The parasympathetic innervation proceeds from S_2-S_4 segments, through the pelvic nerves (McVary et al. 1998; Pennefather et al. 2000; Ali et al. 2004; Powell et al. 2005).

The portion of the major pelvic ganglion innervating the prostate is known as the prostatic plexus; it is located around the prostate capsule and penetrates into the gland through the lateral fascicles of the endopelvic fascia.

The parasympathetic preganglionic neurons innervating the prostate have their perikarya located in the sacral segment of the spinal cord, and the postganglionic neurons are grouped, forming clusters disseminated into the prostate capsule. Thus the capsule is occupied by a tridimensional network of both sympathetic and parasympathetic postganglionic fibers. These fibers go through the capsule to the parenchyma, innervating the acini, the stromal smooth cells, and the blood vessels (Cabo Tamargo et al. 1985; Powell et al. 2005).

The most innervated prostate region is the central zone, with abundant nerve fibers running in parallel to the anterior surface of the seminal vesicles toward the caudal prostate (Benoit et al. 1994). Most of those fibers are distributed in the periacinar stroma, accompanying the smooth fibers and the microvessels (Cabo Tamargo et al. 1985; Benoit et al. 1994; Gkonos et al. 1995a).

Parasympathetic stimuli increase prostate secretion (Ventura et al. 2002), whereas sympathetic action, mediated by α_1-receptors, causes the contraction of prostate smooth muscle and, subsequently, the expulsion of the prostate fluid to the urethral lumen (Chapple et al. 1991; Kurimoto et al. 1998; Mottet et al. 1999).

Immunohistochemical techniques have shown a relevant peptidergic innervation for the prostate (Polak and Bloom 1983; Gu et al. 1983; Crowe et al. 1991; Tainio 1995; Gkonos et al. 1995a; Iwata et al. 2001). The neuropeptides are bioactive polypeptides of variable size that exert their function through interactions with specific receptors, controlling or modulating cell functions by transduction mechanisms that implicate systems of second messengers.

The neuropeptides at the level of the peripheral nervous system might operate either as modulators of the classic neurotransmitters or directly through specific receptors (Hokfelt et al. 1980; Hokfelt 1991). Some neuropeptides are found in relevant amounts in prostate innervation, such as tyrosine neuropeptide (NPY) (Adrian et al. 1984) and vasoactive intestinal peptide (VIP) (Alm et al. 1980; Gu et al. 1983); other peptides have also been observed: calcitonin gene-related peptide (CGRP), substance P, somatostatin, leu-enkephalin, met-enkephalin, histidine-isoleucine (Gu et al. 1983; Fahrenkrug et al. 1989; Lange and Unger 1990; Gkonos et al. 1995a; Tainio 1995).

The function of prostate neuropeptides it is not yet well known; they might exert trophic or neuromodulating actions (Hokfelt 1991; Martin et al. 2000). The evidence for peptidergic fibers around the blood vessels also suggests their intervention in the regulation of local microcirculation (Crowe et al. 1991). Sensory fibers immunoreactive for capsaicin have been also localized in human prostate (Dinis et al. 2005).

2.3
Histology of the Prostate Epithelium

Each prostate lobe is divided into lobules, furnished with excretory ducts draining separately to the prostate urethra. Each of these ducts exhibits subdivisions, giving several branches that connect with secretory units of the tubo-alveolar type (although, frequently these secretory units are called acini). According to topographic criteria, three types of glands can be described in human prostate (Fig. 2):

– Mucosal glands. These are located around the urethra (periurethral glands), are few in number, and constitute the glandular component of the adenomatous nodules in prostate hyperplasia.

– Submucosal glands. Located in a ring external to the periurethral glands.

– Principal glands. Situated more externally to the submucosal layer, these are the most abundant, occupying the rest of the parenchyma.

The mucosal glands drain separately alongside the urethral lumen, whereas both submucosal and principal glands join with the urethra, grouped at the posterior side of the urethral sinus. The prostate epithelium is columnar pseudostratified and has four types of cells (McNeal 1997) (Fig. 3a–h), that are described below.

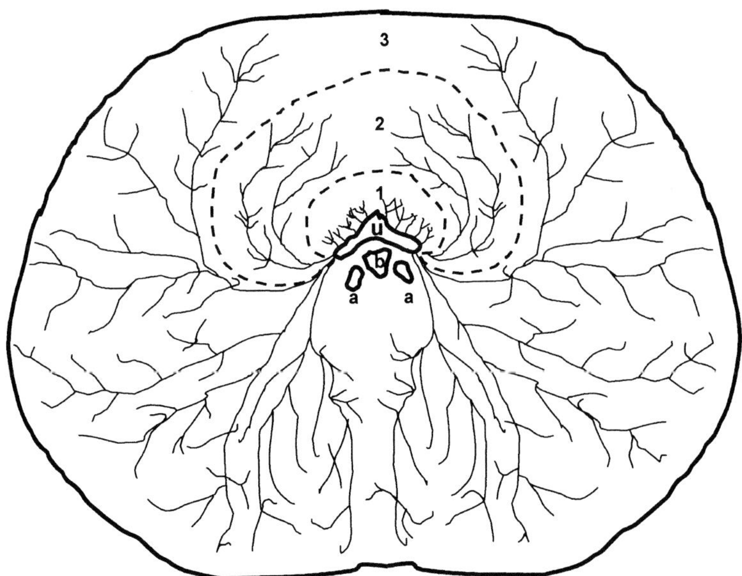

Fig. 2 Scheme of a transverse section of the prostate, showing the topographical types of the glands in human prostate. (*a*) Ejaculatory ducts; (*b*) prostate utriculum; (*u*) urethral lumen; (*1*) mucosal glands; (*2*) submucosal glands; (*3*) principal glands

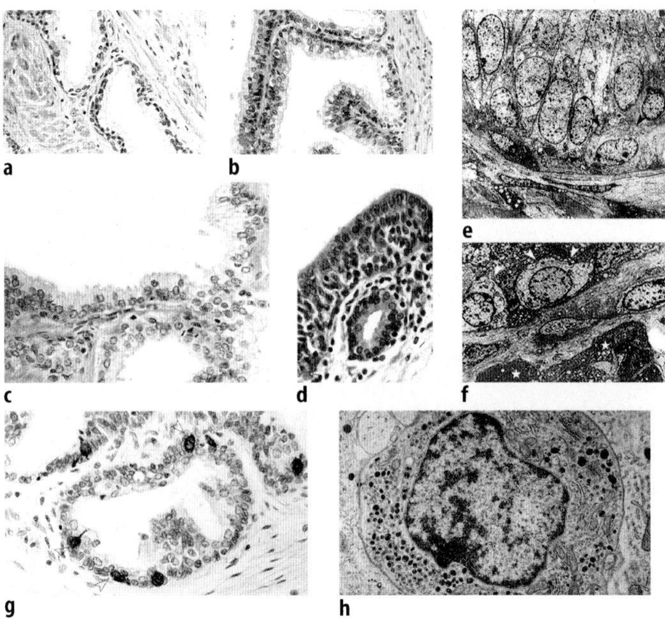

Fig. 3a–h Histology and cell types of the human prostate. **a** Image of gland epithelium and fibro-muscular stroma from the transition zone. The glands are lined by a low columnar epithelium; some basal cells are seen; the stroma is abundant in smooth muscle cells. **b** Image from the central zone. The epithelium shows more pseudostratification than in transition and peripheral zones and more papillary disposition; the basal cell layer is more evident. **c** Histology of peripheral zone. The clear apical cytoplasm of the secretory cells is evident. **d** Transitional epithelium lining the glandular ducts from periurethral region; this is pseudostratified, with abundant cell layers, **a–d** Hematoxylin-eosin. **a, b** ×200. **c, d** ×400. **e** Electron microscopic image of the prostate epithelium. The columnar cells show a clear cytoplasm with basal nuclei and secretory vesicles near the lumen; a cubic basal cell (*arrowheads*) is seen. **f** More details of the ultrastructure of basal epithelium. A basal cell with rounded nucleus and clear cytoplasm (*white arrowheads*) is shown. In the stroma subjacent to the acinus, two contractile cells are seen (*white stars*). **e, f** ×1,500. **g** Neuroendocrine cells (*empty arrowheads*) immunostained to serotonin, in prostate acini from transition zone. ×400. **h** Electron microscopic image from a neuroendocrine prostate cell. Abundant vesicles with electron-dense content are seen distributed around the nucleus. ×3,000

2.3.1
Types of Epithelial Cells

2.3.1.1
Secretory Cells

These are the predominant type in the epithelial layer and they have a columnar shape, with clear cytoplasm (Fig. 3b,c,e). These cells show ultrastructural features consistent with their secretory character: abundant rough endoplasmic reticulum at the basal border, remarkable supranuclear Golgi complex, and secretory vesicles

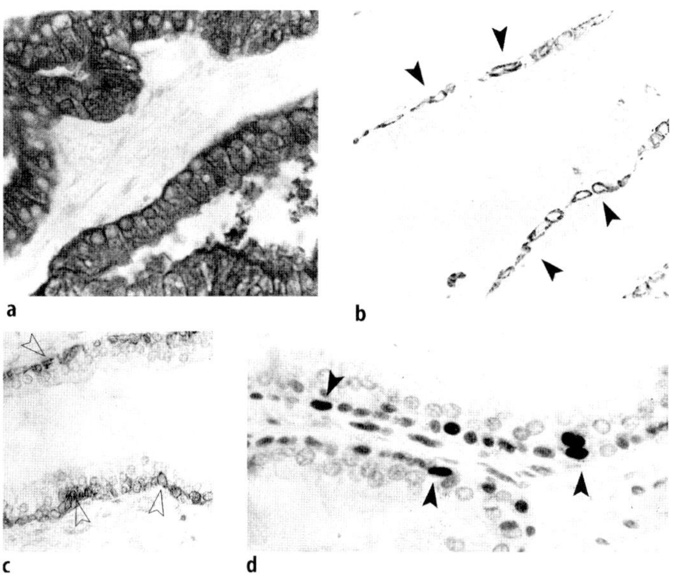

Fig. 4a–d Immunohistochemical features of the prostate epithelium. **a** Prostate epithelium from transition zone immunostained by PSA. **b** Immunostaining by cytokeratins of high molecular weight; only the basal layer (*arrowheads*) shows immunoreactivity. **c** Immunostaining by Bcl-2 protein. Granular immunoreactivity was observed in the basal cells (*empty arrowheads*). **d** Immunostaining by proliferating nuclear antigen (PCNA). Only the proliferating layer (basal cells) show immunoreactivity (*arrowheads*). **a–d** ×400

in the apical border (Fig. 3e). They produce a number of components of the seminal plasma, such as the prostate-specific antigen (PSA) (Fig. 4a) and the prostate acid phosphatase (PAP). All of the prostate zones have a similar pattern of synthesis for both PSA and PAP, but the central zone has a more specific secretion for other proteolytic substances such as pepsinogen II (Reese et al. 1986) and the activating factor of the tissue-type plasminogen (Reese et al. 1988). There are a few morphologic differences between the epithelium of the central zone (granular cytoplasm, more nuclear pseudostratification, and papillary disposition of the epithelial lining) and the epithelium of both transition and peripheral zones (Fig. 3a–c). It is unclear if these slight differences might be attributed to their diverse embryologic origin or to functional particularities (Reese et al. 1988; McNeal 1997).

2.3.1.2
Basal Cells

These are in contact with the basal membrane of the epithelial layer their apical borders do not reach the lumen of the gland, and they are cuboidal in shape, with rounded nuclei and scanty cytoplasm (Mao and Angrist 1966), (Fig. 3e,f). The basal cells show immunoreactivity to the cytokeratins of high molecular weight

(Fig. 4b) (Brawer et al. 1985) but not to PSA or PAP (McNeal 1997); they also show immunoreactivity to Bcl-2 protein (Fig. 4c). This might indicate resistance to apoptosis; in addition, these cells show a remarkable proliferative activity (Fig. 4d).

2.3.1.3
Transitional Cells

The so-called transitional epithelium lines the glandular ducts in their urethral endings. It is pseudostratified, with a variable number of cell layers in relation to the degree of the ductal dilatation. The transitional cells have scanty cytoplasm, and they are isolated from the lumen by a stratum of secretory cells (Fig. 3d) (McNeal 1997).

2.3.1.4
Neuroendocrine Cells

These constitute a prostate cell subpopulation whose origin and role are still being debated; they will be studied in detail later on. The neuroendocrine cells are located in the gland epithelium, among the secretory cells, and they can be identified by immunohistochemical and ultrastructural methods (di Sant'Agnese and Mesy Jensen 1984a, 1987; di Sant'Agnese et al. 1985; Aprikian et al. 1993; Algaba et al. 1995; di Sant'Agnese 1998; Martin et al. 2000; Santamaria et al. 2002) (Fig. 3g,h).

2.4
Benign Prostate Hyperplasia

Benign prostate hyperplasia (BPH) is the most frequent benign disorder of the human prostate (Barry 1990; Kirby 1992). Prostate hyperplasia produces urological symptoms in 54% of males in the age range from 60 to 70 years (Hunter et al. 1996). The frequency of BPH detected in male autopsies by microscopic examination is higher than the frequency of clinical symptoms (around 70% of males in the same age range) (Birkhoff 1983; Berry et al. 1984).

The natural history of BPH shows two stages: pathological and clinical. The pathological stage has a first step with only histologic evidence of several foci of prostate hyperplasia and a second step of macroscopic BPH with evidence of adenomas. About 80% of men above 40 years old will develop microscopic BPH, but only half of this population will present macroscopic BPH in 5 years (Oesterling 1996; Peinado-Ibarra 1998; Fitzpatrick 2006).

The first changes consist of the appearance of stromal nodules around the periurethral glands. The acinar (epithelial) hyperplasia starts close to the stromal nodules and grows slowly over several years (Berry et al. 1984). The transition and periurethral zones are the main regions affected by BPH (Di Silverio et al. 1993; Algaba 1993) (Fig. 5).

The etiopathogenesis of BPH is not yet clearly established (Peehl 1996). There are several theories based on histologic, hormonal, and aging changes, but there

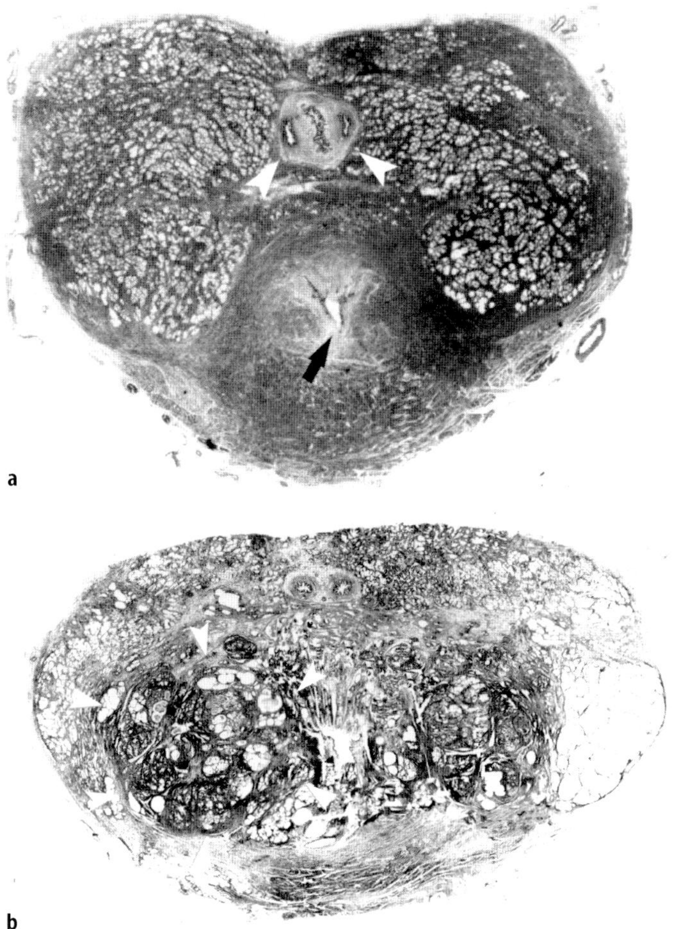

Fig. 5a, b Low-magnification images from transverse sections of normal and BPH prostate.
a Normal prostate from a young adult. The prostate utriculum and ejaculatory ducts (*white
arrowheads*) are surrounded by the central zone glands; the urethral lumen (*arrow*) is
surrounded by stroma of transition zone poor in acinar structures. **b** Prostate from a 70-
year-old man with BPH. A remarkable hyperplastic nodule (*white arrowheads*) is seen near
the urethral lumen. **a, b** H&E, ×4

are no univocal explanations for the origin and development of BPH (Partin et al.
1991). It is possible that the androgenic stimulus serves as a trigger for the stromal
hyperplasia, and the stromal change might induce the epithelial growth (Birkhoff
1983; Narayan 1992).

It has been suggested that neuroendocrine cells are implicated in the genesis of
BPH, perhaps exerting some mediation between stroma and epithelium (Bonkhoff
et al. 1991; Cockett et al. 1993; Algaba 1995; Islam et al. 2002; Rumpold et al. 2002a,b;
Untergasser et al. 2005).

3
The Neuroendocrine System and the Human Prostate

L. Santamaría, L. Alonso

3.1
Introduction

The neuroendocrine cell (NEC) has been revealed as an element of significant importance in the development of proliferative pathologies in the prostate, either BPH or cancer (Untergasser et al. 2005; Slovin 2006). Some historical data about these cells are described in the next paragraphs.

The first observations related to the NECs were reported by Heildenhain (Heildenhain 1870), who demonstrated the presence of a population of chromaffin cells in the gastrointestinal tract, with a possible endocrine role. Later, Masson (Masson 1914) and Hamperl (Hamperl 1932), employing argentic impregnation, demonstrated that these chromaffin cells were also argentaffins, having a wide distribution in a number of organic systems (DeLellis and Dayal 1997).

In 1938, Feyrter (Feyrter 1938) described for the first time the concept of a "diffuse epithelial endocrine system," indicating that argentaffin cells from the gastrointestinal tract might have a local (paracrine) function. Frolich (Frolich 1949) included in this group argentaffin cells discovered in the bronchial epithelium (DeLellis and Dayal 1997). The first findings of NECs in the urogenital tract were reported by Feyrter (urinary bladder) (Feyrter 1938) and Pretl (Pretl 1944), who in 1944 described the presence of argyrophilic cells in the urethro-prostatic region (di Sant'Agnese and Cockett 1994; DeLellis and Dayal 1997).

In 1969 Pearse elaborated the concept of the APUD system (amine precursor uptake and decarboxylation) as a collection of endocrine cells with multiple tissular localization and presenting common functional and biochemical features (Pearse 1969): (a) high content of fluorogenic amines (catecholamines, 5-hydroxytryptamine); (b) ability for the intake of amine precursors (DOPA, 5-hydroxytryptophan); and (c) presence of decarboxylase for the conversion of precursor amino acids into amines.

The APUD system included cells located in "classic" endocrine organs such as pituitary, pineal gland, hypothalamus, suprarenal, parathyroids, etc., and also the so-called chromaffin-argentaffin cells from Langerhans islets, gut, and respiratory tract (Pearse 1969). Furthermore, the term "amine synthesis" was changed to "regulatory peptide synthesis" in the concept of the APUD system (DeLellis and Dayal 1997). Fujita and Kobayashi then used the term "paraneurons" to define NECs with structural, functional, and metabolic features similar to neurons (Fujita et al. 1988). These cells can produce substances like neurohormones or neurotransmitters (Fujita 1977, 1989; Fujita et al. 1980; Luca 1998; DeLellis 2001). The terms endocrine-paracrine cells, neuroendocrine cells, APUD cells, and paraneurons are now employed as synonyms (di Sant'Agnese and Cockett 1994; Montuenga et al. 2003; Toni 2004).

3.1.1
Systemic Distribution of the Neuroendocrine Cells

The neuroendocrine cells are located in a wide variety of organs and systems. They have been described in the central nervous system (hypothalamus, pituitary, pineal gland) (Scharrer 1967; Snyder 1980), respiratory tract (pulmonary neuroepithelial bodies) (Lauweryns and Peuskens 1972; Van Lommel et al. 1999), gastrointestinal tract (Heildenhain 1870; Pearse 1969; di Sant'Agnese and Cockett 1994), thyroid gland (C-cells secreting calcitonin) (DeLellis and Wolfe 1981), skin (Merkel cells) (Gould et al. 1985; Moll et al. 2005; Sidhu et al. 2005), breast (Bussolati et al. 1985; McCarty and Nath 1997), and the urogenital system (Fetissof et al. 1983; Polak and Bloom 1983; di Sant'Agnese and Mesy Jensen 1984a; Fahrenkrug et al. 1989; DeLellis and Dayal 1997).

The NECs from the urogenital tract were first described by Feyrter (1938) and Pretl (1944) in urinary bladder and prostate, by argyrophilic methods. Fetissof confirmed these findings in a variety of urothelial epithelia (Fetissof et al. 1983). The neuroendocrine cells are widely represented in the urethro-prostatic region (di Sant'Agnese et al. 1985; Abrahamsson et al. 1986; di Sant'Agnese and Cockett 1994; di Sant'Agnese 1996; Santamaria et al. 2002).

3.1.2
Embryologic Origin

The embryologic origin of all the NECs is not yet clearly established. The neuroectodermic origin (neural crest) seems to be well founded for some of the neuroendocrine cell populations, such as those from adrenal medulla or from sympathetic ganglia (Le Douarin and Kalcheim 1999). Nevertheless, for other NECs (gastrointestinal and respiratory tracts, pancreas) the endodermic origin is probable (DeLellis et al. 1986; Le Douarin and Dupin 2003; Le Douarin et al. 2004).

The hypothesis of a local origin of the prostate neuroendocrine cells is now well established, because the presence was postulated of a stem cell located in the basal layer of the prostate epithelium that could differentiate to both exocrine and neuroendocrine phenotypes, depending on the action of local signaling agents (Noordzij et al. 1995; Gkonos et al. 1995a). There are several data supporting this hypothesis, including the presence of a number of substances coexpressed in both neuroendocrine and exocrine cell types, such as PSA and serotonin in some NECs, or some neuropeptides, such as NPY in both epithelial columnar cells and neuroendocrine cells (Aprikian et al. 1993; Martin et al. 2000).

3.1.3
Morphological and Histochemical Features of the Neuroendocrine Cells

The NECs are not easily observed with routine stains, such as hematoxylin-eosin. They are oval, pyramidal, or bottle shaped, with clear cytoplasm that contains scarce eosinophilic granules occasionally (DeLellis and Dayal 1997).

Some neuroendocrine cells (gastrointestinal tract, adrenal gland) are brown stained after tissue fixation with chromic acid or potassium dichromate, because of the oxidation of catecholamine contents (chromaffin reaction) (DeLellis and Dayal 1997). Green fluorescence induced by exposure to aldehydes can be also observed in several populations of NECs (Falck and Owman 1965). The neuroendocrine cells also show ability to reduce silver compounds, precipitating metallic silver in their cytoplasm (argentaffin or argyrophilic reaction) (Aguirre et al. 1984).

From a histologic point of view, the neuroendocrine cells are intermingled with other cell types (namely epithelial cells), either isolated or forming small clusters. They lie on the basal membrane of the epithelium (Serezhin 1988), either in contact with the lumen (open type) or not (closed type). Frequently, "dendritic" processes from the NECs can be detected contacting neighboring epithelial secretory cells (Fig. 6); this finding suggests a possible paracrine or regulatory local function (di Sant'Agnese and Mesy Jensen 1984a; Gkonos et al. 1995a; DeLellis and Dayal 1997). Some neuroendings can be occasionally observed contacting neuroendocrine cells from skin, bronchial tree (DeLellis and Dayal 1997), and prostate (Fig. 6) (Santamaria et al. 2002).

The ultrastructure of the neuroendocrine cells indicates their neurocrine features: a remarkable rough endoplasmic reticulum, abundant ribosomes and secretory vesicles (DeLellis and Dayal 1997; Young et al. 2006). These secretory vesicles are the most typical elements of the NECs, with their diameters ranging between 50 and 500 nm and oriented toward the basal pole of the cell (open type) (di Sant'Agnese and Mesy Jensen 1984a; DeLellis and Dayal 1997), whereas in the cells of the closed type the distribution of the secretory granules is more homogeneous (di Sant'Agnese and Mesy Jensen 1984a). Concentric lamellar structures, so-called lamellar bodies, have been observed in some neuroendocrine cells of the prostate. Their significance is uncertain; they would be able to indicate the presence of crinophagic processes—digestion of the excess granules of secretion (di Sant'Agnese and Mesy Jensen 1984a; di Sant'Agnese and Cockett 1994).

The use of immunohistochemistry (Polak and Van Noorden 1988) has evidenced a great variety of both hormonal and nonhormonal products stored in the neuroendocrine cells (Algaba 1995; DeLellis and Dayal 1997). Frequently, one neuroendocrine cell is able to produce more than one peptidic product (di Sant'Agnese and Mesy Jensen 1984b; DeLellis and Dayal 1997). A number of substances (most of them neuropeptides) have been identified in relation to the NEC: gastrin, secretin, cholecystokinin, serotonin, enteroglucagon, somatostatin, substance P, bombesin, inhibitor gastric polypeptide, motilin, pancreatic polypeptide, neuropeptide Y (NPY), vasoactive intestinal polypeptide (VIP), calcitonin gene-related peptide, (CGRP), neurotensin, calcitonin, β-endorphin, synaptophysin, chromogranin, and protein gene product 9.5 (PGP 9.5) (Gould et al. 1986; Polak 1989; di Sant'Agnese 1996).

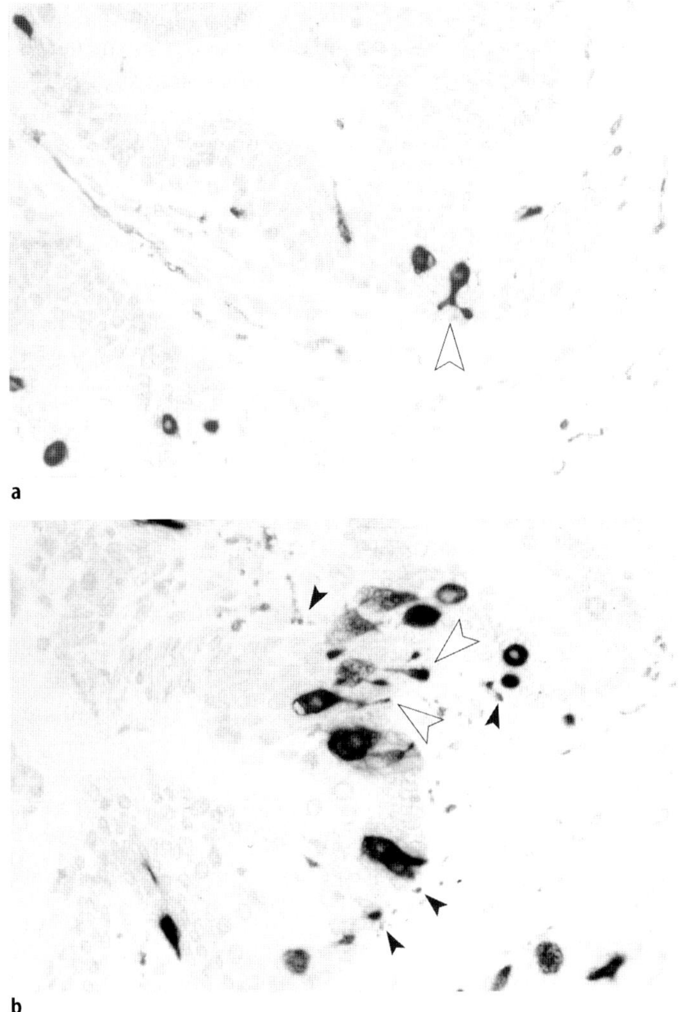

Fig. 6a, b Double immunostaining by chromogranin and PGP 9.5. **a, b** The chromogranin-immunoreactive neuroendocrine cells show frequently dendritic-like processes at the basal pole (*empty arrowheads*). **b** Neuroendings immunostained by PGP 9.5 are seen in contact with neuroendocrine cells (*black arrowheads*). **a, b** ×400

3.2
Neuroendocrine Cells in the Normal Prostate

3.2.1
Distribution

The neuroendocrine cells are widely distributed in the urethro-prostatic region, prostate glandular ducts, and peripheral zone of the prostate (di Sant'Agnese et al.

1985; Abrahamsson et al. 1986; Cohen et al. 1993c; di Sant'Agnese and Cockett 1994; Noordzij et al. 1995), whereas seminal vesicles and ejaculatory ducts are devoid of NECs (di Sant'Agnese et al. 1985; Aprikian et al. 1993). Cohen (Cohen et al. 1993) found neuroendocrine cells in the prostate stroma, but this observation has been not confirmed in other studies.

3.2.2
Natural History

There are few studies about the evolution of the neuroendocrine cell population in human prostate because it is not easy to obtain representative samples of all the developmental stages from normal human prostate. However, knowledge of distribution, changes in number with age, and normal histophysiology of the NECs is very important to ascertain their behavior in pathologic conditions of the prostate.

In this sense, one of the studies that stand out is that of Cohen (Cohen et al. 1993), in which a total of 63 prostates from autopsies of individuals whose ages ranged between 1 month and 70 years were studied. The authors analyzed the zonal distribution and observed that neuroendocrine cells were more abundant at the transition zone (including periurethral zone) than in the ducts and peripheral glands. NECs were present in all the regions in the newborn and remained up to 3 months of age. Afterwards, they decreased in the peripheral zone, reappeared later in puberty, and rose to the highest number at 50 to 70 years of age. NECs from the transition zone remained unchanged throughout life. These findings suggested a neuroendocrine cell population dependent on androgenic stimulus in the peripheral zone and another population independent of androgens in the transitional zone, therefore suggesting the possibility of two functional types of neuroendocrine cells (Cohen et al. 1993).

Some authors have observed a decrease in neuroendocrine cells in the peripheral zone from the fifth decade of life. This decrease has not been related to other significant changes, such as the presence of foci of prostate intraepithelial neoplasia (PIN) or prostatic carcinoma (Algaba et al. 1995).

3.2.3
Expression of Androgenic Receptors

Some initial immunohistochemical studies found positive expression in the majority of neuroendocrine cells in normal prostate and prostate cancer (Nakada et al. 1993). However, subsequent studies did not confirm this fact (Krijnen et al. 1993; Bonkhoff et al. 1993; Iwamura et al. 1994a; di Sant'Agnese 1996).

It is now thought that androgens do not exert a direct regulating effect on the NEC. Nevertheless, the existence of some kind of indirect regulation seems probable. This would explain, according to some authors, the differences found by Cohen in the distribution of neuroendocrine cells relating to age (Cohen et al. 1993; di Sant'Agnese and Cockett 1994). Neuroendocrine cells would be able to secrete their products independently of the regulation exercised by the androgens, and they would theoretically be able to continue secreting them during androgenic suppression (di Sant'Agnese 1995, 2001; Tilley et al. 1996; Bentel and Tilley 1996; Evangelou et al. 2004; Sciarra et al. 2006).

3.2.4
Expression of Epidermal Growth Factor

Immunohistochemistry has demonstrated that the neuroendocrine cells of the normal prostate express immunoreactivity to receptors of epidermal growth factor (EGF) (Iwamura et al. 1994b). EGF is a mitogenic factor that plays an important role in the growth and differentiation of a great variety of tissues and cell lines (Fisher and Lakshmanan 1990; Sherwood and Lee 1995).

Some studies seem to indicate the existence of a reciprocal regulation between EGF and parathyroid hormone-related peptide (PTHrP) (Ferrari et al. 1992; Alsat et al. 1993). This fact is interesting, because expression of PTHrP was demonstrated in prostate neuroendocrine cells (Iwamura et al. 1993, 1994c).

3.2.5
Physiology of Prostate Neuroendocrine Cells

The role of the neuroendocrine cell in the prostate is not yet well known, but the hypothesis that it is implicated in the growth, differentiation, and regulation of prostate is plausible (di Sant'Agnese 1992, 1996, 1998; di Sant'Agnese and Cockett 1994; Algaba and Trias 1995; Noordzij et al. 1995; Gkonos et al. 1995a). This hypothesis is based on three characteristics, according to di Sant'Agnese:

1. The cell morphology: Dendritic processes of NECs extend themselves between the surrounding epithelial cells, which suggests that the neuroendocrine cell could exercise paracrine regulation.
2. The secretion of products of known biological activity, identified by means of immunohistochemical techniques.
3. The analogy with other elements of the diffuse neuroendocrine system or APUD system (di Sant'Agnese and Cockett 1994).

Some authors suggest that NECs would be able to play a role in human fertility, because they are abundant in the prostatic utriculus (Guy et al. 1998).

It is possible that neuroendocrine cells secrete products toward the stroma and have receptors for stromal factors that would provide the necessary interactions for the normal growth and physiology of the prostate (di Sant'Agnese and Cockett 1994; Herrero et al. 2002; Montuenga et al. 2003).

3.2.6
Quantification of Neuroendocrine Cells

There are several studies on quantification of the number of neuroendocrine cells (di Sant'Agnese et al. 1987; Cohen et al. 1993; Algaba et al. 1995), but generally these are based on counting of cell profiles, either per unit of area or per microscopic field; these methods are not suitable for a true, unbiased estimation of the number of cells, and they can lead to important systematic errors that affect the validity of the study. If we bear in mind that the prostate and its different cell subpopulations are three-dimensional, it is only adequate to express the quantification either as the number of cells per unit of volume (numerical density) or as the number of cells per prostate total volume (Howard and Reed 2005a).

Stereological methods allow the performance of unbiased estimates of the number of cells contained in a volume. In normal prostate, the estimates of absolute and relative (numerical density) numbers of neuroendocrine cells immunoreactive to chromogranin A, PGP 9.5, and serotonin (Santamaria et al. 2002) (Fig. 7) agree with previous semiquantitative observations (Cohen et al. 1993; Noordzij et al. 1995; Laczko et al. 2005) of a predominance of neuroendocrine cells in the transition zone of the normal prostate (Fig. 8). The concordance between absolute and relative high numbers of neuroendocrine cells in the transition zone indicates that the number of these cells was volume independent, because the contribution of the transition zone to prostate volume is only approximately 5% (McNeal 1997). Thus the important population of neuroendocrine cells observed in the transition zone could be related to the hypothesis formulated by Aumuller (Aumuller et al. 1999): During early stages of prostate development, the number of neuroendocrine cells in the urethral epithelium is considerably increased, and these cells are transported into the gland with the continuous sprouting and lumen formation of prostate anlagen. Thus the farther the acini are located from the periurethral region (i.e., for those located in peripheral zone), the more diluted the neuroendocrine population. Other hypotheses could be alternatives to the transport hypothesis to explain the scarcity of neuroendocrine elements found in the central zone (Fig. 8): Although the Wolffian origin of some components of human prostate is still controversial (Aumuller et al. 1999; Laczko et al. 2005), some authors (McNeal 1990) assign a Wolffian embryogenic origin to the central zone of the prostate, and it is well known that the Wolffian derivatives lack neuroendocrine cells (Santamaria et al. 1993; Aumuller et al. 1999; Sommerfeld et al. 2002).

The transition zone is the prostate region where BPH originates (McNeal 1997), and it has been established that neuropeptides and possibly serotonin can activate prostate growth (Cockett et al. 1993; Golomb et al. 2000; Jongsma et al. 2000a). Thus the abundant neuroendocrine cells of that region could play a role in the increase of cell proliferation (Meyer et al. 1982) observed in both epithelial and stromal compartments of hyperplastic nodules. The relative increase in neuroendocrine cells was reported in small, early-developed hyperplastic nodules (Cockett et al. 1993). Nevertheless, some other studies (Noordzij et al. 1995; Martin et al. 2000) described a significant decrease in the global population of chromogranin-immunoreactive neuroendocrine cells in BPH specimens (Fig. 9). This decrease can be explained by the fact that the specimens studied came from large, hyperplastic nodules from patients with long-term BPH (Martin et al. 2000).

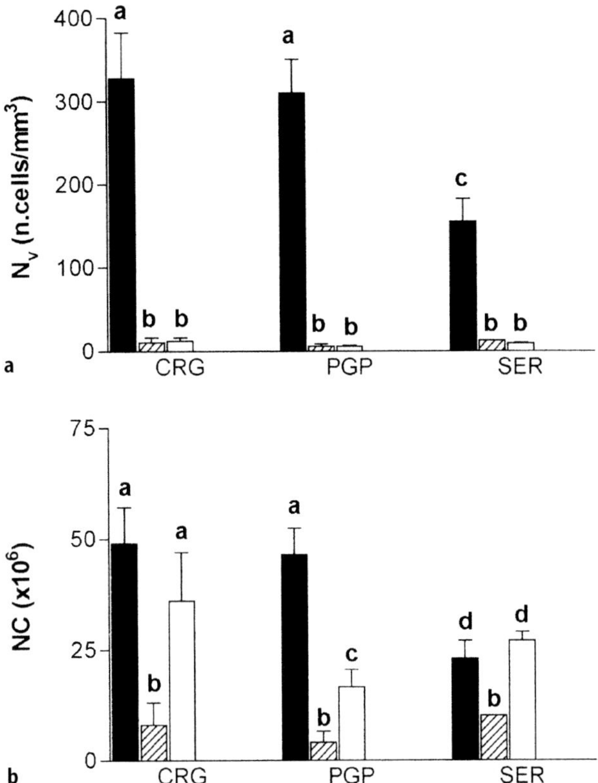

Fig. 7a, b Bar diagrams indicating means ± SD. **a** Numerical density (N_V) of neuroendocrine cells immunostained to chromogranin A (*CRG*), PGP 9.5 (*PGP*), and serotonin (*SER*) expressed in number of immunoreactive cells x $10^3/mm^3$ of epithelial tissue from transition zone (*black bars*), central zone (*striped bars*), and peripheral zone (*empty bars*). **b** Absolute number (NC) of neuroendocrine cells immunostained to chromogranin A (*CRG*), PGP 9.5 (*PGP*), and serotonin (*SER*) expressed in millions of immunoreactive cells from transition zone (*black bars*), central zone (*striped bars*), and peripheral zone (*empty bars*). In each graph, the *letters* over each *bar* indicate the significance: *Bars* labeled by different *letters* show significant differences (*P*<0.05). (From Santamaría et al. 2002)

Fig. 8a–f Regional variations of immunostained neuroendocrine cells in normal prostate. **a** Double chromogranin A-PGP 9.5 immunostaining shows abundant neuroendocrine cells in transition zone (*TZ*). **b** This region also shows a remarkable number of serotonin immunoreactive cells, but in lesser numbers than the chromogranin-positive cells observed in **a**. **c** Double chromogranin A-PGP 9.5 immunostaining shows a lesser number of neuroendocrine cells in central zone (*CZ*). **d** The scarcity of serotonin-immunostained cells is also evident in *CZ*. **e** Double chromogranin A-PGP 9.5 immunostaining shows a low number of neuroendocrine cells in peripheral zone (*PZ*). **d** The scarcity of serotonin-immunostained cells is also evident in *PZ*. **a–f** ×400

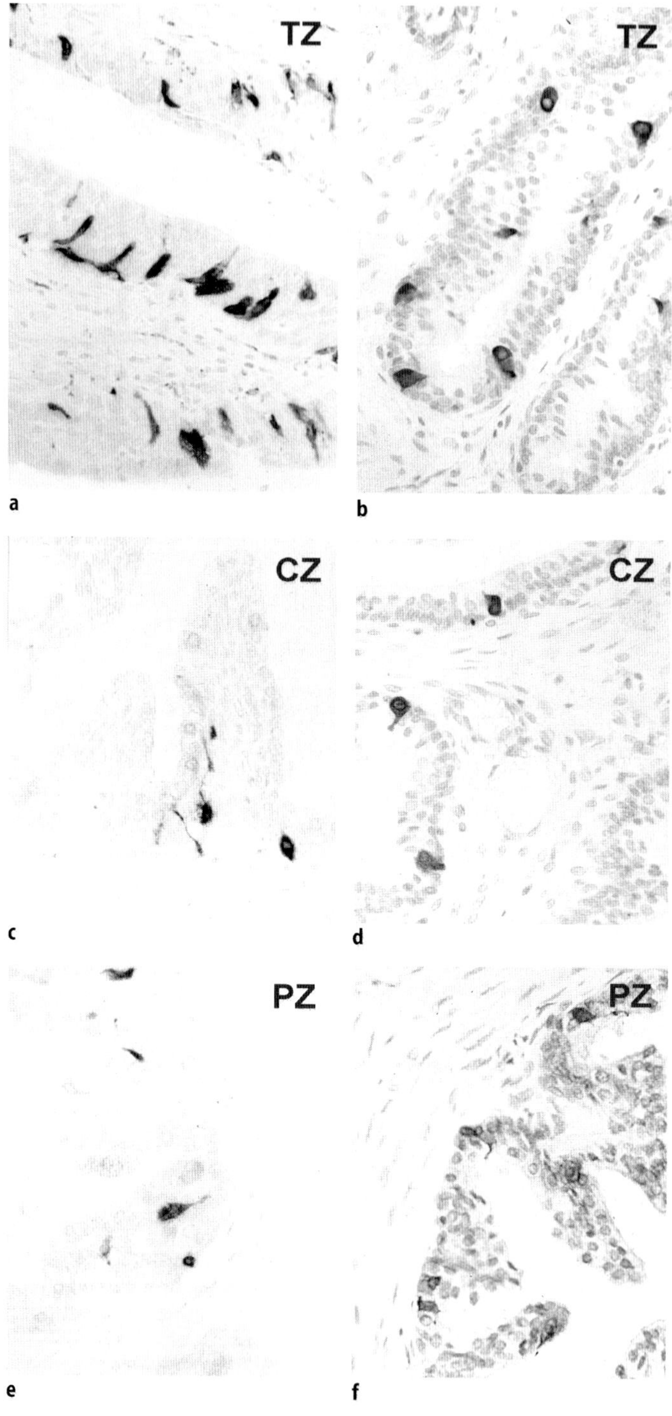
TZ
TZ
CZ
CZ
PZ
PZ
a
b
c
d
e
f

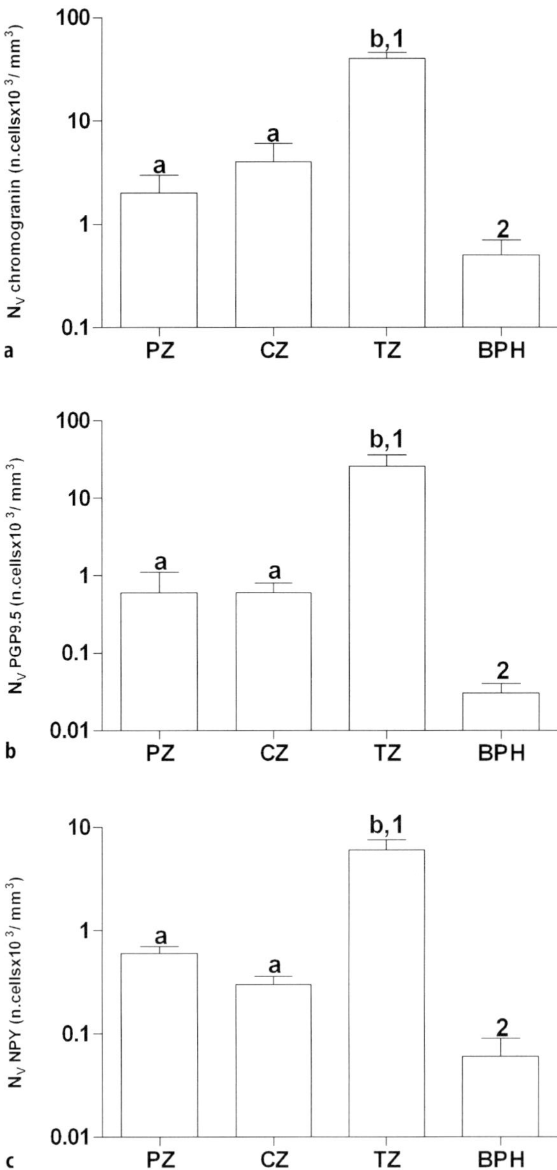

Fig. 9a–c Bar diagrams indicating means ± SD. Numerical density (N_V) of neuroendocrine cells immunostained against chromogranin A (**a**), PGP 9.5 (**b**), and NPY (**c**) expressed in number of immunoreactive cells × 10^3/mm³ of epithelial tissue from peripheral zone (*PZ*), central zone (*CZ*), transition zone (*TZ*), and benign hyperplasia specimens (*BPH*). For comparison among prostate regions the *letters* over each *bar* indicate the significance, and for comparison between TZ of normal prostate and BPH specimens the *numbers* over each *bar* indicate the significance. *Bars* labeled by different *letters* or *numbers* show significant differences ($P<0.05$). (Modified from a table in Martin et al. 2000)

The absolute and relative numbers of serotonin-immunoreactive cells in the transition zone represent approximately 50% of the number of chromogranin- or PGP 9.5-immunostained cells, a figure in agreement with the fact that not all neuroendocrine cells express serotonin. Other neuropeptides produced by prostate neuroendocrine cells have been detected: CGRP, calcitonin, katacalcin, somatostatin, and so forth (di Sant'Agnese and Mesy Jensen 1984b; di Sant'Agnese et al. 1989; Abrahamsson et al. 2000). It is also interesting to note that the peripheral zone (the site where prostate adenocarcinoma most frequently originates) shows, in absolute terms, a number of serotoninergic neuroendocrine cells as high as the transition zone; thus the significant presence of neuroendocrine cells secreting neuropeptides in the peripheral zone could be correlated with the induction of androgen-independent growth in prostate carcinoma (Jongsma et al. 2000a). Although the question of whether neuroendocrine differentiation in prostatic cancer is associated with poor prognosis remains controversial (Ahlegren et al. 2000; Ito et al. 2001; Kollermann and Helpap 2001), it has been reported that neuroendocrine cells may be related to the development of androgen-independent tumors (Jongsma et al. 2000b; Amorino and Parsons 2004). Moreover, recent insights into the expression of antiapoptotic proteins by neuroendocrine cells of the prostate (Xing et al. 2001) provide a potential molecular basis for the persistence of these cells in refractory prostate carcinoma.

In disagreement with other authors (Aumuller et al. 1999), we found that the number of neuroendocrine cells immunoreactive to PGP 9.5 was considerably higher in normal prostate, with special reference to the transition zone. This discordance may be explained by the fact that the study by Aumuller et al. (Aumuller et al. 1999) did not indicate whether the specimens studied came from the transition zone of adult prostate.

3.3
Neuroendocrine Markers in the Prostate

The most common neuroendocrine markers are summarized in Table 1. They can be grouped as nonneuropeptidic and neuropeptidic markers: The term neuropeptidic indicates the presence of an active peptide with a putative neurotransmitter function, whereas a nonneuropeptidic marker can have an active (hormonal) role, for example, serotonin, or another functional/structural role (chromogranin, synaptophysin). All these markers can be localized in neuroendocrine cells, nerve fibers, or both.

3.3.1
Nonhormonal Markers
3.3.1.1
Chromogranins

The chromogranins are proteins that constitute the principal components of the secretion granules of the NECs (Vittoria et al. 1990; di Sant'Agnese 1996; DeLellis

Table 1 Neuroendocrine and peptidergic innervation markers in human prostate

	NECs	Innervation	References
Nonneuropeptidic markers			
Chromogranin	Yes	No	Schmid et al. 1994
			Gkonos et al. 1995a
			di Sant'Agnese 1996
Neuron-specific enolase	Yes	No	Haimoto 1985
Synaptophysin	Yes	No	Gould et al. 1986
PGP 9.5	Yes	Yes	Rode et al. 1985
			Santamaría et al. 1993
Serotonin	Yes	No	Cockett et al. 1993
			di Sant'Agnese 1996
Neuropeptidic markers			
CGRP	Yes	Yes	di Sant'Agnese 1989
			Chapple et al. 1991
			Gkonos et al. 1995a
NPY	Yes	Yes	Higgins and Gosling 1989
			Gkonos et al. 1995a
			Martin et al 2000
VIP	No	Yes	Higgins and Gosling 1989
			Gkonos et al. 1995a
SP	No	Yes	Gu et al. 1983
			Gkonos et al. 1995a
Somatostatin	Yes	Yes	Gu et al. 1983
			di Sant'Agnese
			and De Mesy 1984
			Gkonos et al. 1995a
			di Sant'Agnese 1996
TRH/TSH	Yes	Yes	Abrahamsson et al. 1986, 1989
			Gkonos et al. 1995a
PTHrP	Yes	No	Gkonos et al. 1995a

Developmental stages are used in the embryonic period

and Dayal 1997). There are three principal types: chromogranin A, chromogranin B, and chromogranin C (secretogranin II) (Feldman and Eiden 2003). The function of these proteins is not clearly known, but it seems to be related to processing and storage of regulating peptides (DeLellis and Dayal 1997). The chromogranins are distributed throughout all of the neuroendocrine system, predominating as one of three types in a particular tissue; for example, in pancreatic neuroendocrine cells chromogranin A predominates, with little expression of chromogranin B (Hagn et al. 1986).

The majority of neuroendocrine cells from prostate express immunoreactivity to chromogranins (Iwamura et al. 1994a); therefore the chromogranins are useful as a marker for these cells in both normal and pathologic glands (Abrahamsson et al. 1989; Theodoropoulos et al. 2005). Although the three chromogranins are expressed in the normal prostate and in nodular hyperplasia (Schmid et al. 1994), chromogranin A is more frequently identified (Gkonos et al. 1995a). In cases of carcinoma with neuroendocrine differentiation, the predominant type was chromogranin B (Schmid et al. 1994).

3.3.1.2
Neuron-Specific Enolase

This is an enzyme that was initially employed as a marker of nerve cells, smooth muscle fibers, and neuroendocrine cells. Today it is considered to be of low specificity, since it has been found in a great variety of tissues: lymphocytes, tubular renal cells, type II pneumocytes, epithelial bronchial cells, etc. (Haimoto et al. 1985).

3.3.1.3
Synaptophysin

This is a glycoprotein integrated in the membranes of synaptic vesicles. It has been demonstrated in a wide range of cells and employed in the diagnosis of neuroendocrine tumors (Gould et al. 1986). It is believed that the role of synaptophysins can be related to the intracellular transportation and liberation of hormones and neurotransmitters (Portela-Gomes et al. 1999).

3.3.1.4
Protein Gene Product 9.5

Protein gene product 9.5 (PGP 9.5) is a protein of 27,000 kDa with ubiquitin carboxyl-terminal hydrolase activity (Wilkinson et al. 1989), initially isolated from human brain. By means of immunohistochemical methods, PGP 9.5 has been demonstrated in the nervous system and in the neuroendocrine cells from different species of mammals (Rode et al. 1985; Santamaria et al. 1993; Martin et al. 1995; Edyvane et al. 1995). Also, it was localized in nonneuroendocrine cells, like Leydig cells, ovarian follicles, tubular renal cells, and glioma cells (Giambanco et al. 1991), and in epididymis of rat and human (Santamaria et al. 1993; Martin et al. 1995).

3.3.1.5
Serotonin

This is a biogenic amine very frequently found in the NECs (di Sant'Agnese 1996). Serotonin is a neurotransmitter, a vasoactive agent, and a factor that stimulates the proliferation of fibroblasts and smooth muscle (Cockett et al. 1993; Noordzij et al. 1995; Dizeyi et al. 2004). Serotonin might play an important role in prostate cell proliferation and could be implicated in the development of prostate pathology (Cockett et al. 1993).

3.3.2
Neuropeptidic Markers

3.3.2.1
Calcitonin and Calcitonin Gene-Related Peptide

Calcitonin gene-related peptide (CGRP) was characterized in 1982, and it results from alternative processing of the RNA transcript of the calcitonin gene. This alternative processing gives two 37-amino acid peptides, α- and β-CGRP, that differ by one amino acid (Amara et al. 1982). CGRP is a vasoactive agent, mediating sensitivity to pain and modulating muscular tone (Amara et al. 1982; Gkonos et al. 1995a). Some authors observed immunoreactivity to CGRP in a subpopulation of neuroendocrine cells from the human urethro-prostatic region, and the location of the calcitonin secretory granules was described by electron microscopy (di Sant'Agnese et al. 1989; Abrahamsson et al. 2000). CGRP-positive nerve fibers were detected in the human prostate (Chapple et al. 1991); its distribution was parallel to that observed for substance P, but the CGRP-positive neuroendings were of lesser abundance than those of substance P (Fahrenkrug et al. 1989). CGRP's actions in the prostate seem to be related to smooth muscle contractility (Gkonos et al. 1995a).

3.3.2.2
Neuropeptide Y

Neuropeptide Y (NPY) is formed from 36 amino acids, and it is characterized by having terminal tyrosine groups. Initially, it was extracted from porcine brain (Adrian et al. 1984; Fahrenkrug et al. 1989). It has similarity to the pancreatic polypeptide and to the YY peptide (Aguilar et al. 2004). NPY is distributed extensively within the autonomous nervous system (Crowe et al. 1991). In the urogenital tract, it has been found in sympathetic neuroendings (Ekblad et al. 1984; Chapple et al. 1991; Iwasa 1993; Gkonos et al. 1995a; Edyvane et al. 1995). NPY-immunoreactive nerve fibers are frequently distributed around arteries of medium and large caliber, on nonvascular smooth muscle, and also in the vicinity of epithelial cells (Higgins and Gosling 1989; Fahrenkrug et al. 1989). NPY fibers are especially abundant in seminal vesicles, where they form thick nets around the epithelial lining (Lange and Unger 1990). In the ureterovesical region, NPY fibers are preferentially distributed in the distal ureter and in the vesical trigonum (Smet et al. 1994; Edyvane et al. 1995). Although NECs positive for NPY were not initially described in the prostate, we reported their presence in prostate epithelium from all the prostate zones, colocalizing with PGP 9.5 and chromogranin A (Martin et al. 2000) (Fig. 10). According to some authors, NPY would be able to act by modulating either sympathetic contractility or the effects of other neurotransmitters on epithelial cells (Ekblad et al. 1984; Gkonos et al. 1995b). A relationship between cholinergic and NPYergic fibers has been detected in a study performed in monkey prostate, and the authors have suggested that

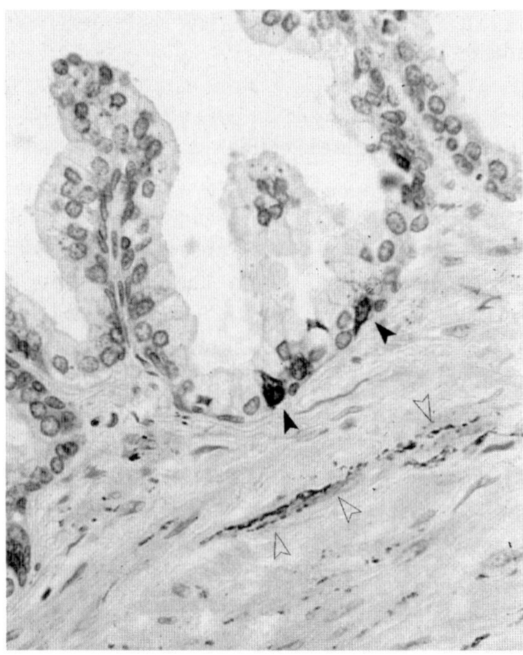

Fig. 10 Peripheral zone from a normal prostate immunostained by NPY. Two neuroendocrine cells show NPY immunoreactivity (*black arrowheads*). Some nerve fibers immunostained by NPY were also seen (*empty arrowheads*). ×400

NPY fibers could regulate cell secretion by acting upon the cholinergic system, or else have a direct effect on the secretory cells (Yokoyama et al. 1990).

3.3.2.3
Vasoactive Intestinal Polypeptide

Vasoactive intestinal polypeptide (VIP) is a 28-amino acid polypeptide. It was originally isolated in porcine intestine and recognized as having a powerful vasodilator effect (Fahrenkrug et al. 1989). In the urogenital tract, the fibers immunoreactive to VIP are intimately associated with blood vessels (mainly arteries and arterioles), not vascular smooth muscle and glands (Larsen et al. 1981; Fahrenkrug et al. 1989). VIP-immunoreactive nerves are abundant in the penis's erectile tissue. Also, they appear in great number in the seminal vesicles, deferent duct, epididymis, urinary bladder, ureters, and prostate (Vaalasti et al. 1980; Gu et al. 1983; Polak and Bloom 1984; Lange and Unger 1990; Edyvane et al. 1995). VIP fibers frequently appear surrounding prostatic acini (Higgins and Gosling 1989; Crowe et al. 1991). Some authors have found colocalization of VIP and nitric oxide synthetase immunoreactivity in prostate neuroendings. Both substances have been related to the control of muscular tone (Smet et al. 1994; Hedlund et al. 1997). VIP has also been associated with cholinergic fibers (Gkonos et al. 1995a). Experimental studies in the rat prove that epithelial prostatic cells contain receptors to VIP, and its stimulation produces an increase in both cAMP and glandular secretion (Carmena and Prieto 1985; Solano et al. 1996; Gutierrez-Canas et al. 2005). It is possible that VIP exercises a regulatory role on the secretory function of the epithelial cells in the human prostate (Gkonos et al. 1995a). A possible role in the relaxation of vascular smooth muscle and in the increase of prostatic blood flow has also been attributed to VIP (Larsen et al. 1981; Fahrenkrug J et al. 1989; Edyvane et al. 1995). Neuroendocrine cells immunoreactive to VIP have not been described in the prostate.

3.3.2.4
Substance P

The peptide substance P (SP) is comprised of 11 amino acids. It has been found in very limited quantities in the urogenital tract (Gu et al. 1983). The nerve fibers immunoreactive to SP have a similar distribution to that observed for CGRP fibers, and they are located in close relation to smooth vascular and nonvascular muscle and to epithelial cells (Fahrenkrug et al. 1989). Among the functions attributed to substance P are vasodilation, contraction of smooth muscle, glandular secretion, and regulation of the immune response (Hokfelt 1991); its role in the prostate is unknown.

3.3.2.5
Somatostatin

This hypothalamic peptide is comprised of 14 amino acids; it was isolated and characterized chemically in 1973. It is widely distributed within the central and peripheral nervous systems (Vittoria et al. 1990; Barnett 2003). Somatostatin is an inhibitor of the release of multiple pituitary hormones. At the extrapituitary level, gastrointestinal activities have also been described , such as inhibition of insulin release and release of other peptide hormones, decrease of visceral blood flow, and inhibition of gastric and duodenal motility. In seminal vesicles and human prostate, several authors have identified somatostatin-positive fibers (Gu et al. 1983; di Sant'Agnese and Mesy Jensen 1984b; Crowe et al. 1991; Jen and Dixon 1995; Gkonos et al. 1995a; Tainio 1995; di Sant'Agnese 2001; Mosca et al. 2004). Somatostatin fibers innervate the smooth muscle and the glandular interstitium of prostate, as well as the smooth muscle of the seminal vesicles and deferent ducts (Gu et al. 1983; Tainio 1995). Several authors suggest that somatostatin inhibits the exocrine secretion of the glandular epithelium (Vittoria et al. 1990).

3.3.2.6
Peptides Related to Thyrotropin-Releasing Hormone and to Thyrotropin

Thyrotropin-releasing hormone (TRH) has been isolated principally in the central nervous system. The existence of peptides related to TRH was also demonstrated in several other sites: rat pancreas (Kawano et al. 1983), rat prostate (Pekary et al. 1983), and human prostate and seminal fluids (Cockle et al. 1994). The role of these peptides in the human prostate is unknown. It is thought that they can regulate the actions of other peptides, like thyrotropin (TSH)-related peptides, and they have also been found in prostate (Gkonos et al. 1995a; Tainio 1995; di Sant'Agnese 1996). Some authors have related increased levels of TRH-related peptides to the occurrence of BPH (Cockle et al. 1994).

3.3.2.7
Parathyroid Hormone-Related Peptide

The peptide parathyroid hormone-related peptide (PTHrP) was initially found in the paraneoplastic syndrome associated with hypercalcemia (Heath et al. 1990; Iwamura et al. 1993). Subsequent studies demonstrated PTHrP expression in the normal tissue of several organs like brain, intestine, adrenal gland, skin, and placenta (Iwamura et al. 1994d). PTHrP expression has been demonstrated in a subpopulation of neuroendocrine cells, in both human normal prostate and BPH, and this finding suggests that they should play a role in growth and glandular differentiation (di Sant'Agnese 1995), as well as in the regulation of calcium in the seminal fluid (Iwamura et al. 1994d).

3.4
Neuroendocrine Cells in Benign Prostate Hyperplasia

The neuroendocrine cell may play a role in the pathogenesis of BPH, as an intermediary between epithelium and stroma. The NEC would receive certain stromal signals of a biochemical type and would become active, secreting, in turn, other factors that would induce the proliferation of the epithelial cells (Algaba et al. 1995). This supposition gets stronger if it is considered that neuroendocrine cells are often near the epithelial proliferating cells, although they do not seem to proliferate (Bonkhoff et al. 1991; Gkonos et al. 1995a).

Serotonin, expressed by the majority of neuroendocrine cells (di Sant'Agnese et al. 1985; Abrahamsson et al. 1986; Dizeyi et al. 2004), regulates the proliferation of the epithelial cells and stimulates the mitogenesis of fibroblasts and smooth muscle, and this molecule has been implicated in the pathogenesis of BPH (Cockett et al. 1993). Cockett compared the number of NECs in normal prostate tissues and in BPH specimens. Quantification of 5-hydroxytryptamine (5-HT) and its metabolites was also performed by HPLC. A marked decrease of cells in the majority of the big hyperplastic nodules, as compared with normal prostatic tissue, was ascertained. These findings were corroborated with quantitative analysis that showed a significant decrease in 5-HT levels in BPH, as compared with the normal prostate. Similar results were observed with the use of stereological nonbiased methods to estimate the number of NECs. The numerical density of neuroendocrine cells immunoreactive to chromogranin A and to PGP 9.5 was significantly decreased in BPH nodules compared with the transition zone of normal prostate (Fig. 9) (Martin et al. 2000). However, in recently formed hyperplastic nodules , with remarkable epithelial cell proliferation, a greater number of NECs was found. It can be concluded that neuroendocrine cells would be able to play a role in early BPH pathogenesis by local regulatory mechanisms (Cockett et al. 1993). Davis et al. corroborate this observation, detecting a high concentration of serotonin in tissular homogenates from hyperplastic nodules in initial stages of BPH, as compared with those of advanced stages (Davis et al. 1989).

4
Innervation and Neuroendocrine Cells in Both Normal and Hyperplastic Prostate

L. Santamaría, L. Alonso

4.1
Neuroendocrine Markers in Normal Prostate and BPH

Besides the observation of NECs positive for chromogranin, PGP 9.5, serotonin, and NPY (Fig. 8) in the normal prostate, we have found occasional neuroendocrine cells that express synaptophysin at the transition zone (Fig. 11a), whereas some cells expressing arginine vasopressin (Fig. 11b) have been detected in the central zone. Neuroendocrine cells immunoreactive to somatostatin and CGRP were also detected in the transition zone (Fig. 11c,d).

Neuroendocrine cells were also observed in BPH specimens immunostained for chromogranin, PGP 9.5, and serotonin (Fig. 12a,b), although the number of these cells was low in comparison to that quantified in normal prostate (Fig. 9). Occasionally neuroendocrine cells expressing NPY have been detected (Fig. 12). Nevertheless, synaptophysin-, arginine vasopressin-, somatostatin-, and CGRP-immunostained cells were not found in BPH.

The neuroendocrine cells described in the present study (in the normal prostate and in BPH as well) have morphologic characteristics that other authors have shown: oval, pyramidal or bottle shaped, they are dispersed among the other epithelial cell types. Some have processes toward the glandular lumen, and others do not (Figs. 8, 11, 12). Sometimes, some cells show characteristic processes toward the neighboring epithelial cells. All the above might stress the possible paracrine role of the NECs in the prostate (di Sant'Agnese and Mesy Jensen 1984a; Gould et al. 1986; Abrahamsson et al. 1986; di Sant'Agnese et al. 1989; Aprikian et al. 1993; Cockett et al. 1993; Noordzij et al. 1995; Gkonos et al. 1995a; di Sant'Agnese 1996; DeLellis and Dayal 1997).

By means of double immunostaining methods, cells that simultaneously express immunoreactivity against two antigens have been visualized: chromogranin-serotonin, chromogranin-PGP 9.5, and chromogranin-NPY. This agrees with ultrastructural findings obtained by other authors that colocalize cytoplasmic vesicles

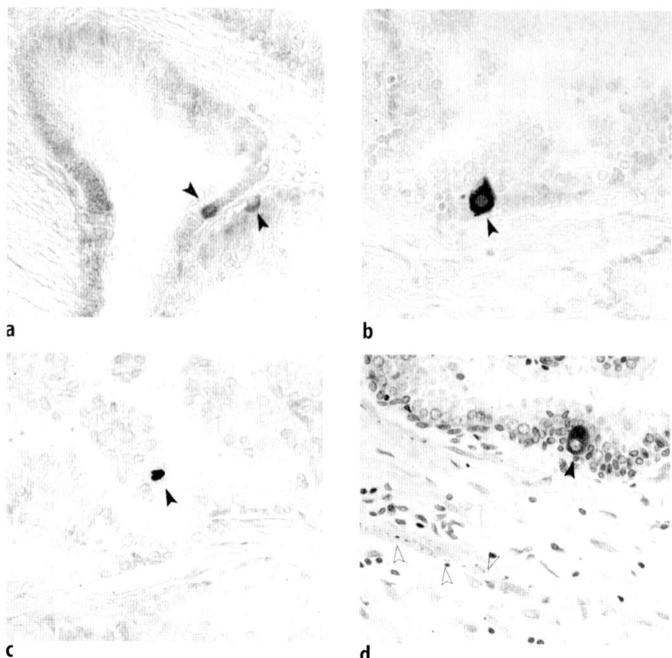

Fig. 11a–d Occasional immunoreactive neuroendocrine cells to synaptophysin are seen in transition zone (*black arrowheads*) (**a**). In central zone a cell positive for arginine vasopressin is observed (*black arrowhead*) (**b**). In transition zone a neuroendocrine cell immunostained by somatostatin (**c**) and CGRP (**d**) was detected (*black arrowheads*). In **d**, nerve fibers immunostained by CGRP (*empty arrowheads*) were also seen. **a–d** ×400

with different size and shape; this suggests that the same neuroendocrine cell may be implicated in the secretion of different substances (di Sant'Agnese and Mesy Jensen 1984b).

We have occasionally observed, in normal specimens and in BPH, expression of NPY in the cytoplasm of the secretory epithelial cells (Fig. 13) (Martin et al. 2000). A possible explanation would be based on the common local origin of neuroendocrine and secretory cells (Noordzij et al. 1995; Gkonos et al. 1995a; Rumpold et al. 2002a).

The expression of somatostatin, CGRP, and synaptophysin in normal NECs agrees with findings described by other authors (Gould et al. 1986; di Sant'Agnese and Cockett 1994b; Gkonos et al. 1995a). We have also observed cells that express arginine vasopressin, although in very limited numbers. This peptide is secreted by the neurohypophysis and implicated in the regulation of plasma osmolarity. This finding has not been described previously in the prostate, and might be related to regulation of the composition of the prostatic secretion fluid.

The absence of neuroendocrine expression of somatostatin, CGRP, synaptophysin, or arginine vasopressin in BPH might indicate that in hyperplasia a selective

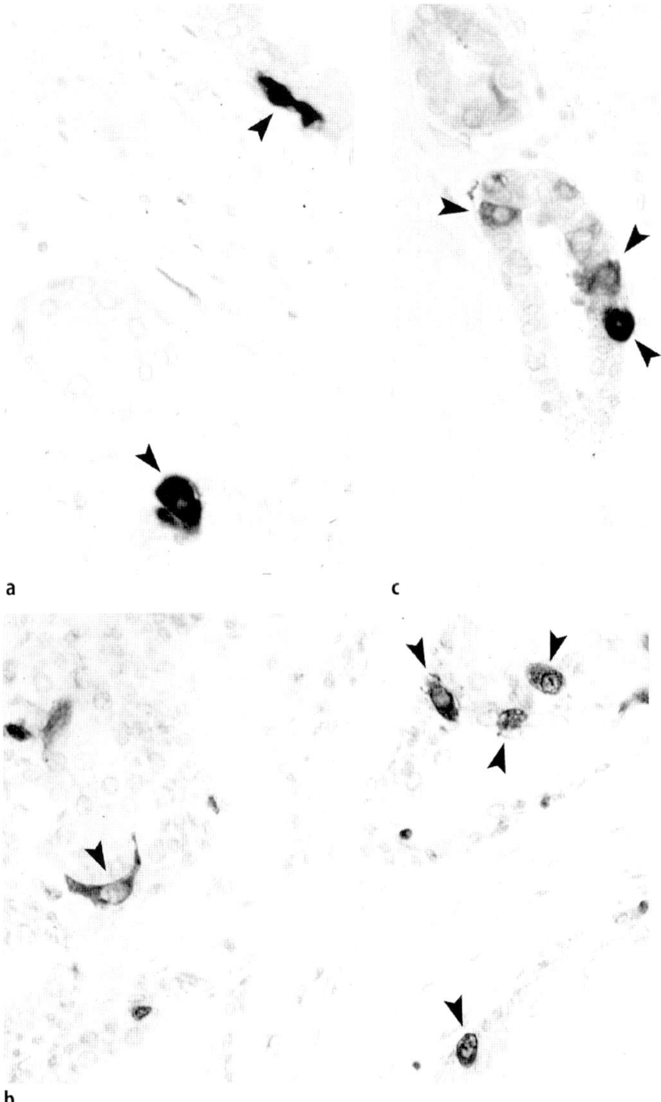

Fig. 12a–c Neuroendocrine cells in BPH. **a** Double immunostaining by chromogranin A-PGP 9.5 shows some cells immunoreactive for both antigens (*black arrowheads*). **b** Several serotonin-immunoreactive cells are detected (*black arrowheads*). **c** Neuroendocrine cells immunostained by NPY are seen (*black arrowheads*). **a–c** ×400

depletion of some types of neuroendocrine cells is produced. As a consequence, in BPH functionally important changes derived from the loss of the regulating neuropeptides would take place in comparison to the physiological conditions that exist in normal prostate.

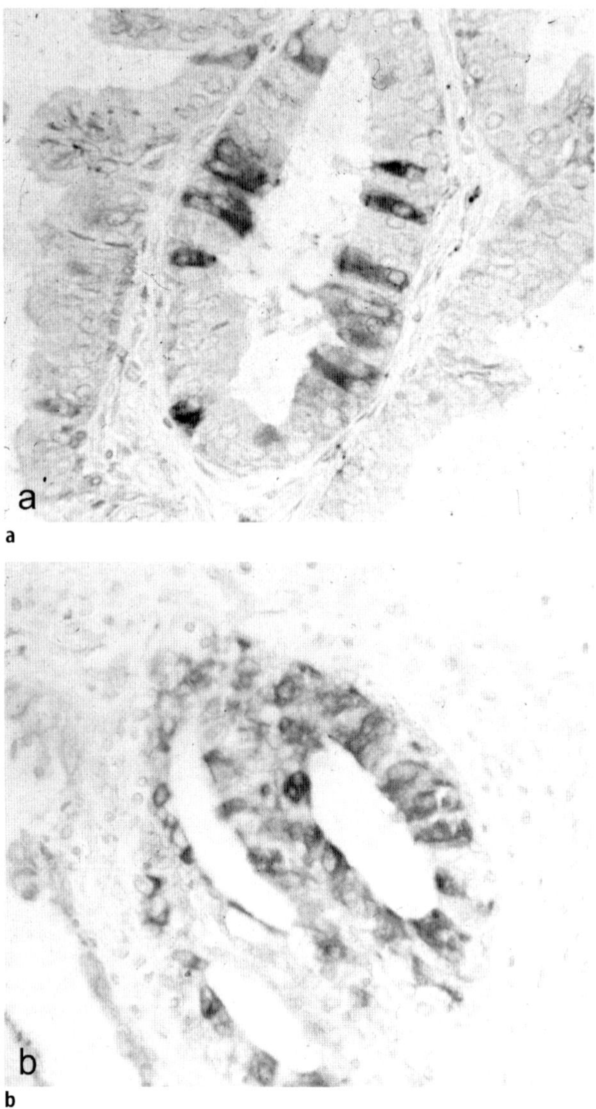

Fig. 13a, b Epithelial secretory (nonneuroendocrine) cells, immunostained by NPY were seen in normal prostate (**a**) and in BPH (**b**). ×400

4.1.1
Neuropeptide Innervation in Normal Prostate and BPH. Qualitative Observations

The distribution of nerve fibers immunoreactive to PGP 9.5 agrees with what some authors have previously described (Jen and Dixon 1995). PGP 9.5-immunoreactive fibers are abundantly located in all prostatic regions of normal specimens, especially in the peripheral region. PGP-positive, thick nerve bundles are not infrequently found in the stroma close to the prostate capsule (Fig. 14a). An abundant network of PGP 9.5-immunoreactive neuroendings is seen around acini in all prostate regions (Fig. 14b).

The distribution of the nerve fibers immunoreactive to NPY and VIP in normal prostate and in BPH was similar from a qualitative point of view to that described by other authors (Gkonos et al. 1995a); they are populating the interglandular stroma and surrounding the prostatic acini. The NPY-positive fibers seem more abundant at the peripheral zone and around the blood vessels (Fig. 15a,b) and more evident in normal cases than in BPH (Fig. 15c,d). VIP fibers were more abundantly observed around the acini and in the periglandular stroma in both normal and hyperplastic glands (Fig. 16a,b).

We have observed limited fibers immunoreactive to CGRP, located at the transition zone (Fig. 11d) The role of this neuropeptide in the human prostate can be related to the modulation of the contractile response of the stromal cells (Gkonos et al. 1995a; Jen and Dixon 1995). On the other hand, CGRP-immunostained fibers have been not detected in BPH.

4.1.2
Neuropeptide Innervation in Normal Prostate and BPH. Quantitative Observations

From a quantitative point of view, the best manner to estimate the number of nerve fibers in a certain tissue is to calculate either relative or absolute length of nerve fibers (Howard and Reed 2005b). Nevertheless, in human specimens, often obtained from surgery or archival material, it is not always possible to meet all the sampling criteria (preservation of isotropy, for example) required to perform these measurements in an unbiased way. Thus other stereological estimates such as volume density of innervation, which does not require such stringent sampling conditions, were carried out with human specimens. The density of volume of innervation (volume fraction, i.e., the ratio between the immunoreactive area of nerve fiber and the total area of reference) of fibers positive for PGP 9.5, NPY, and VIP does not change significantly among regions of normal prostate; nevertheless, a significant decrease in these estimates was seen in BPH as compared with each of the zones of the normal prostate (Fig. 17a–c). It has been said that peptidergic innervation could decrease with aging; this would be particularly remarkable in BPH specimens, since they all derived from older patients. In addition, data observed by other authors (Chow et al. 1997) indicate that adrenergic nerves from the stomach could decrease with aging. There is proof of the linkage between autonomic innervation and the changes that can affect other organs of the lower urinary tract during life. In advanced age, a decrease in the number of neuronal axons and cholinergic nerves takes place in the bladder (Gilpin et al. 1986). It has also been observed that the density of noradrenergic innervation in the urinary tract of the adult rat is limited compared with the innervation in young animals (Warburton and Santer 1994).

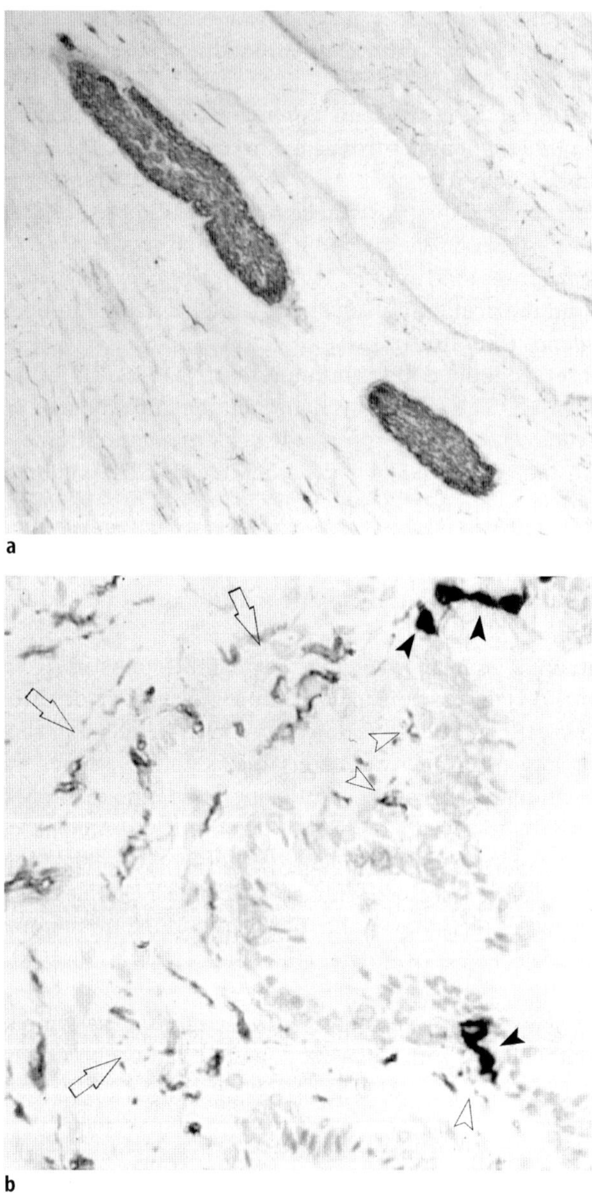

Fig. 14a, b Nerve bundles immunoreactive to PGP 9.5 were observed at subcapsular level in normal prostate (**a**). Transition zone (**b**): Abundant PGP 9.5-immunoreactive nerve fibers are shown in periacinar stroma (*empty arrows*) and near the basal membrane of the acini (*empty arrowheads*); the epithelium also shows PGP 9.5-positive neuroendocrine cells (*black arrowheads*). **a** ×200; **b** ×400

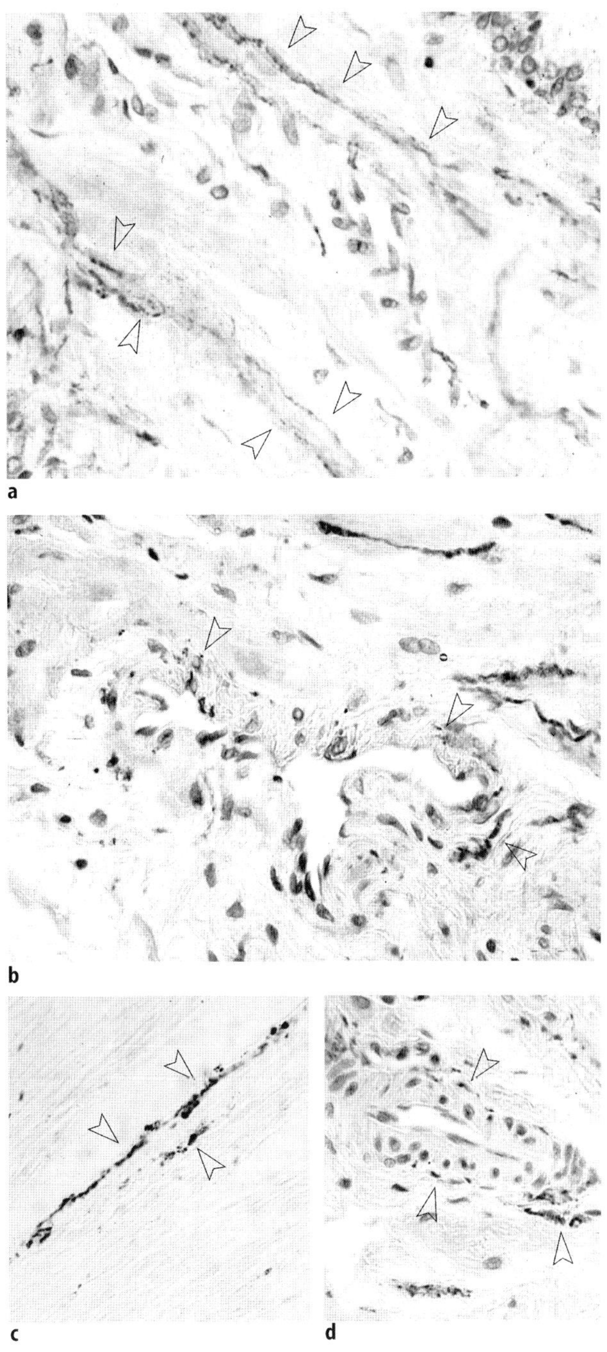

Fig. 15a–d Nerve fibers immunoreactive to NPY in normal and BPH prostate. **a** Abundant NPY-positive fibers (*empty arrowheads*) are seen in the stroma around the acini of peripheral zone. **b** NPY-immunoreactive neuroendings around an arteriolar vessel (*empty arrowheads*) in peripheral zone of normal prostate. **c** NPY fibers in the stroma of a BPH nodule (*empty arrowheads*). **d** NPY-immunoreactive neuroendings around an arteriolar vessel (*empty arrowheads*) in BPH tissue. **a–d** ×400

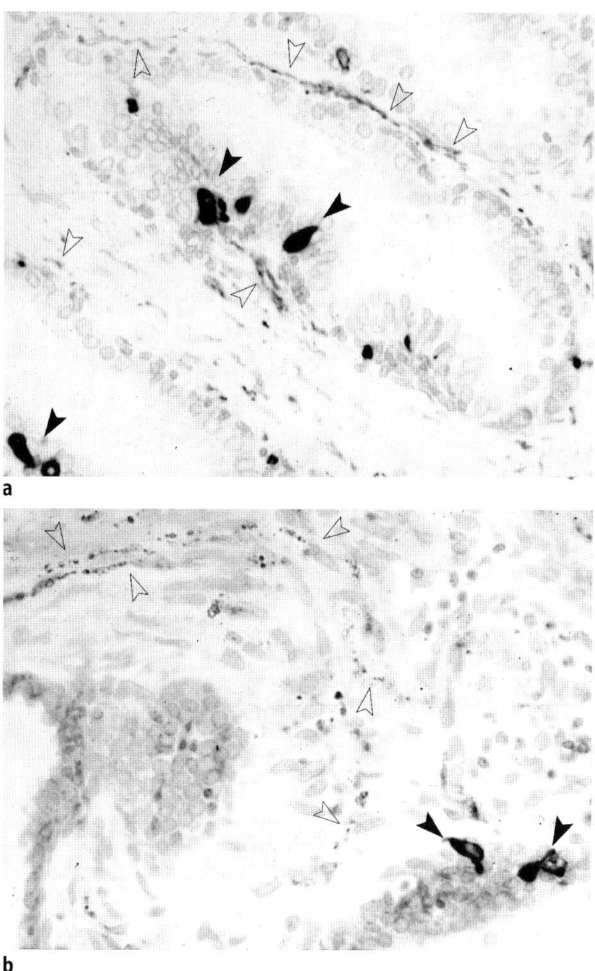

Fig. 16a, b VIP immunostaining in normal and BPH prostate. **a** Double immunostaining by chromogranin A-VIP in transition zone. The neuroendocrine cells are immunostained only by chromogranin A (*black arrowheads*), whereas VIPergic fibers are detected around the acini (*empty arrowheads*). **b** Double immunostaining by Chromogranin A-VIP in a BPH specimen. A few neuroendocrine cells are immunostained only to chromogranin A (*black arrowheads*), whereas abundant VIPergic fibers are detected in the periacinar stroma (*empty arrowheads*). **a, b** ×400

When the absolute volume of innervation (PGP 9.5, NPY, and VIP fibers) was measured, a significant increase in these fibers was observed in the peripheral zone in comparison with central and transition regions and when compared with BPH (Fig. 17d–f).

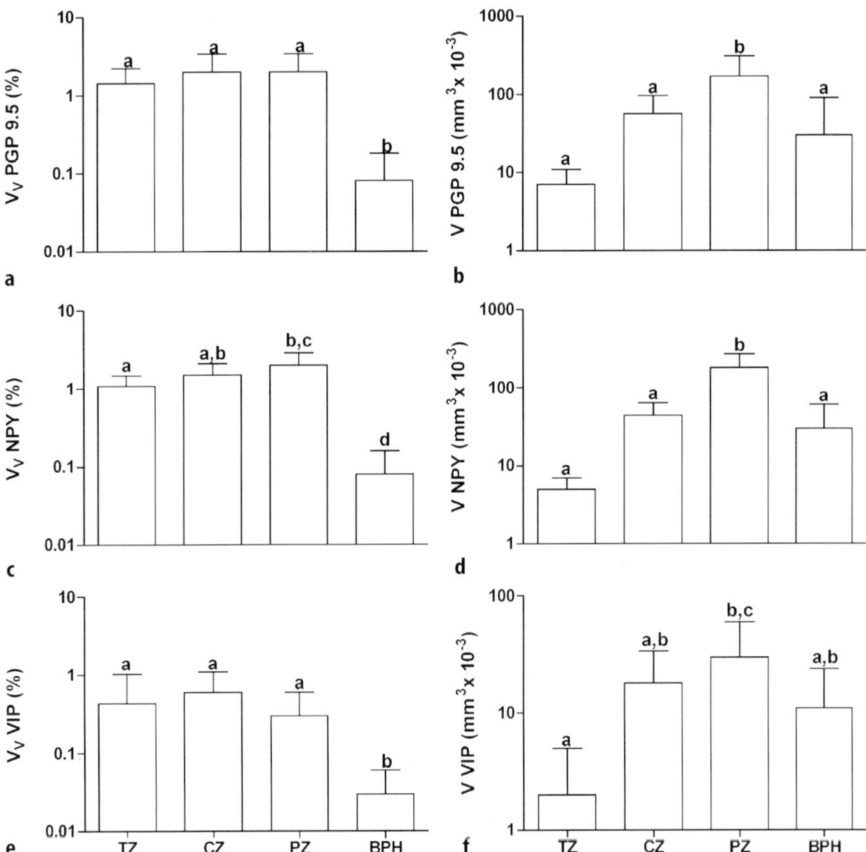

Fig. 17a–f Bar diagrams indicating means ± SD. Volume fraction (V_V) of nerve fibers immunostained to PGP 9.5 (a), NPY (b), and VIP (c) expressed in percentage of immunoreactive area of nerve fibers over the total immunoreactive and nonimmunoreactive area of stromal tissue from transition zone (*TZ*), central zone (*CZ*), peripheral zone (*PZ*), and benign hyperplasia specimens (*BPH*). Total volume (V) of nerve fibers immunostained by PGP 9.5 (d), NPY (e), and VIP (f) expressed in $mm^3 \times 10^{-3}$ of immunoreactive nerve fibers, from transition zone (*TZ*), central zone (*CZ*), peripheral zone (*PZ*), and benign hyperplasia specimens (*BPH*). For comparison among prostate regions and BPH cases the *letters* over each *bar* indicate the significance. *Bars* labeled by different *letters* show significant differences ($P<0.05$)

4.2
Conclusions

Several conclusions can be drawn. A predominance of neuroendocrine cells was observed in the transition zone of the normal prostate. The neuroendocrine cells of that region could play a role in the genesis of BPH. The significant presence of neuroendocrine cells secreting neuropeptides in the peripheral zone could be correlated with the induction of androgen-independent growth in prostate carcinogenesis. The Wolffian origin attributed to the central zone can explain its poor population of neuroendocrine cells. It has been observed that in BPH there is a decrease in the number of neuroendocrine cells that might be selective (those immunoreactive to somatostatin, CGRP, arginine vasopressin, and synaptophysin).

In the normal prostate the region most densely innervated by peptidergic fibers was the peripheral zone; this can be related to the increasing presence of stromal contractile cells in this region. The peptidergic innervation, and more relevantly, the NPY fibers, was significantly decreased in BPH. This might be related either to a general phenomenon of aging or to a specific pathologic change from BPH.

5
The Prostate of the Rat

I. Ingelmo, L. Santamaría

The prostate varies among the different animal species in its anatomy, biochemistry, and physiopathology. The mature prostate of the mammals is a glandular organ comprised of epithelial and stromal cells regulated by a number of hormones (steroids, prolactin) and other molecules (growth factors, neuropeptides, biogenic amines, etc.) (Lee et al. 1997; Hedlund et al. 1997; Reiter et al. 1999).

5.1
Development of the Rat Prostate

The initial event in prostate morphogenesis is the sprouting of epithelial cords (prostatic buds) from the epithelium of the urogenital sinus (endodermal component) toward the mesenchyma (mesodermal component) surrounding the urogenital sinus. In rodents this occurs according to a spatial pattern that establishes the lobar subdivisions of the prostate (Cunha et al. 1987; Timms et al. 1994).

The critical period for ductal sprouting in rodents, and the consequent process of branching and growing, starts around day 15 of gestation and finishes approximately 4–5 weeks after birth (Sugimura et al. 1986; Hayashi et al. 1991). Branching morphogenesis in mice is completed 2 weeks after birth (Sugimura et al. 1986). At this moment the blood levels of testosterone are low, and the increase in prostatic weight is moderated. In puberty, the level of androgens (testosterone) increases significantly and both prostate weight and prostatic DNA content also increase rapidly (Donjacour and Cunha 1988). By contrast, the morphogenesis of the human prostate occurs entirely during the fetal period, with ductal development occurring during the first half of pregnancy (Xue et al. 2001).

Although the rat prostate is an androgen-dependent organ (Huggins and Russel 1946), several studies indicate that it is androgen sensitive but does not require a chronic androgenic stimulus (Donjacour and Cunha 1988).

The prepubertal development of the ductal prostate system is not uniform: The growth is bigger at the distal location than in the portion proximal to the urethra (Cunha et al. 1986; Lee 1996).

5.2
Macroscopic Description of the Rat Prostate

The rat prostate is an exocrine gland structured around the urethra, and unlike the human gland, it presents a lobar organization (Jesik et al. 1982). The prostate of the rat is constituted of three lobes that, according to their position around the proximal region of the urethra, can be designated as ventral, dorsal, and lateral lobes. Each is composed of a complex system of ducts proximally joining to the urethra and distally finishing in many secretory acini divided into branches and placed close to the prostate capsule (Fig. 18). The individual lobes do not encircle the urethra completely (Cunha et al. 1987). This explains why rodents, unlike humans, do not suffer symptoms of the lower urinary tract due to enlargement of the prostate (Maini et al. 1997).

All the components of the prostate are surrounded by a thin connective capsule. Anatomic individualization between the dorsal and lateral lobes is difficult; therefore, these lobes are frequently considered as a single element (dorso-lateral lobe).

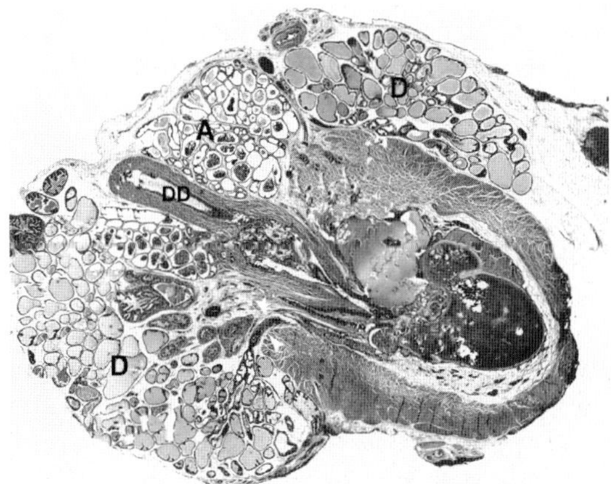

Fig. 18 Transverse section of rat prostate at low magnification. *A*, ampular region; *D*, dorso-lateral region; *DD*, deferent duct. *Arrowheads* indicate the presence of excretory glandular ducts. H&E, ×4

The glandular component of the rat prostate has a histologic structure of the tubo-alveolar type. All the tubo-alveoli from a lobe drain through a number of excretory ducts. The branching pattern of the prostatic ducts exhibits some heterogeneity according to the lobe considered (Hayashi et al. 1991; Timms et al. 1994; Kinbara and Cunha 1996).

The ventral and lateral lobes drain to the urethra by means of two or three principal ducts, which show a so-called "oak tree" branching pattern (Sugimura et al. 1986). The ducts from lateral lobes show two regions: type I, formed by five to seven long ducts spreading cranially toward the seminal vesicles and type II formed by five or six short ducts that branch out caudally to the neck of the urinary bladder. The ventral lobe shows two or three pairs of thin ducts originating in the central portion of every lobe and following a parallel course to the layers of smooth muscle in the wall of the urethra (Hayashi et al. 1991; Timms et al. 1994; Kinbara and Cunha 1996).

The dorsal lobes drain to the urethra with 10–14 narrow ducts (5–6 for each side) having a mixed morphology between acinus and duct. The more posterior acini of the dorsal lobes lead into the medial zone of the urethra by four or five ducts. The remaining acini drain by means of ducts that penetrate into the roof of the dorsal urethra (Hayashi et al. 1991; Timms et al. 1994; Kinbara and Cunha 1996). These ducts have a "palm tree" branching pattern (Sugimura et al. 1986).

5.2.1
Anatomic Relationships of the Prostate

Other structures in intimate anatomic relationships with the rat prostate are: the ampular glands, the seminal vesicles, the coagulating glands, the deferent ducts, the ureters, the urinary bladder, and the urethra.

The ampular glands surrounding the distal portion of the deferent ducts (Fig. 18) can be considered as an integrating part of the prostate because their secretion contributes to the composition of the prostatic fluid. Out of every ampular gland originate from 20 to 30 ducts, draining to the prostate urethra at a point located to the rear of the other lobar ducts.

5.2.2
Prostate Blood Vessels

The arterial flow to the rat prostate comes from branches arising from the iliac internal artery, named prostatic arteries or vesical inferior arteries. These arteries arrive at the gland between the ventral and dorsal lobes (Greene 1955). These arteries form the periurethral arterial circle, from which all arteriolae that irrigate the prostate emerge (Jesik et al. 1982; Scolnik et al. 1992).

The ventral lobe receives vascular irrigation from the artery of the ventral lobe, which originates in the periurethral arterial circle, and from arteries of the adipose-prostatic superior and inferior bundles (Shabsigh et al. 1999).

The dorso-lateral lobe is irrigated from the artery of the lateral lobe, from the periurethral arterial circle, and from arteries traveling in the adipose-prostatic inferior bundle.

The veins that drain the prostatic system emerge from the prostate surface. These veins form the venous prostatic plexus, draining to the iliac internal vein.

The capillary density of the prostate is directly proportional to the size of the acini and remains constant over 90 days of postnatal development (Scolnik et al. 1992).

5.3
Histology of the Rat Prostate

5.3.1
Generalities

The first descriptions of the histological architecture of the rat prostate go back several decades (Gunn and Goud 1957; Flickinger 1972). The prostatic tissue is structured in two compartments, epithelial and stromal, the proportion of the components differing according to species. In the adult rat, the ratio stroma/epithelium is 1/5, while in human prostate the stroma and the epithelial cells are present about in the same proportion (DeKlerk and Coffey 1978; Bartsch and Rohr 1980).

5.3.2
Histology of the Acini

The three pairs of lobes are connected to the urethra by connective tissue and the ductal excretory system. The structure of glandular component is tubo-alveolar, although the secretory units are traditionally designated as acini. The morphology of the epithelial cells that line the acini differs according to the lobe considered and its topographic location (Nemeth and Lee 1996).

The ventral acini are highly contorted structures, especially the peripheral units (near the prostate capsule), which show abundant circumvolutions and are smaller than the central acini (near the urethra). The acini are surrounded by the stroma constituted by loose connective tissue. The luminal secretion of the acini is pale and eosinophilic. The cells of the epithelial lining of the ventral acini show a basophilic cytoplasm, predominantly columnar in shape, intermingled with occasional cuboidal cells with basal nuclei. An irregular stratum of basal cells, located between the basal membrane and the columnar cells, is also observed. The basal cells are cubic or flattened, with an ovoid nucleus and scant cytoplasm that does not reach the acinar lumen (Nemeth and Lee 1996; Arriazu et al. 2005).

The acini from the lateral lobe are bigger than those of the ventral lobe, of variable size, and very convoluted in shape. The acinar secretions are intensely stained with eosin. The epithelial cells are either cuboidal or columnar, containing nuclei located centrally.

The acini from dorsal lobes are ample and less contorted than the ventral and lateral ones and are loosely distributed within the stroma. The luminal secretions present an intermediate staining between those seen with the ventral and lateral lobes. The epithelial cells are cuboidal, with central nuclei.

The ampular acini are lined by a cubic-flattened epithelium. The secretion is characterized by intense eosinophily and the presence of nonstained holes (bubbles) (Fig. 18).

5.3.3
Histology of the Glandular Ducts

The prostate ducts are contorted in shape, and their wall is constituted of remarkable fibromuscular stroma formed by connective tissue and several layers of smooth muscle fibers (Fig. 19a–c). The ductal system presents regional variations related to its morphology and functional activity. Thus the ducts can be divided into regional segments defined as proximal, intermediate, or distal according to their proximity to the urethra (Lee et al. 1990; Nemeth and Lee 1996).

The ductal epithelial cells can be classified, attending to morphologic and functional criteria, in three different types (Lee et al. 1990; Sensibar et al. 1991). The epithelium from the distal ducts is columnar (Fig. 19a) and exhibits apical nuclei with frequent proliferating activity (Lee et al. 1994). The epithelium from the intermediate ducts represents an intermediate stage, being columnar or cubic in shape, with basal nuclei and abundant apical clear cytoplasm (Fig. 19b), indicating secretory activity; these cells are able to synthesize prostate-specific proteins. The epithelial cells from proximal ducts are cubic (Fig. 19c), and many of them experience apoptosis (Kerr and Searle 1973).

At the distal region of the excretory ducts, a discontinuous layer of smooth muscle cells can be observed . The ducts from the intermediate region have one or two continuous muscle layers that at the proximal level are increased to four strata near where it joins the urethra (Nemeth and Lee 1996).

6
Neuroendocrine Cells in the Rat Prostate

R. Rodríguez, J.M. Pozuelo

6.1
Introduction

The neuroendocrine cells (NEC) of the prostate are included in the so-called diffuse neuroendocrine system (Feyrter 1938; Pearse 1969; Montuenga et al. 2003), which is characterized by the synthesis and secretion of polypeptides with biological activity, either locally or through the blood, where they rise to concentrations sufficient to act like circulating hormones (DeLellis and Dayal 1997).

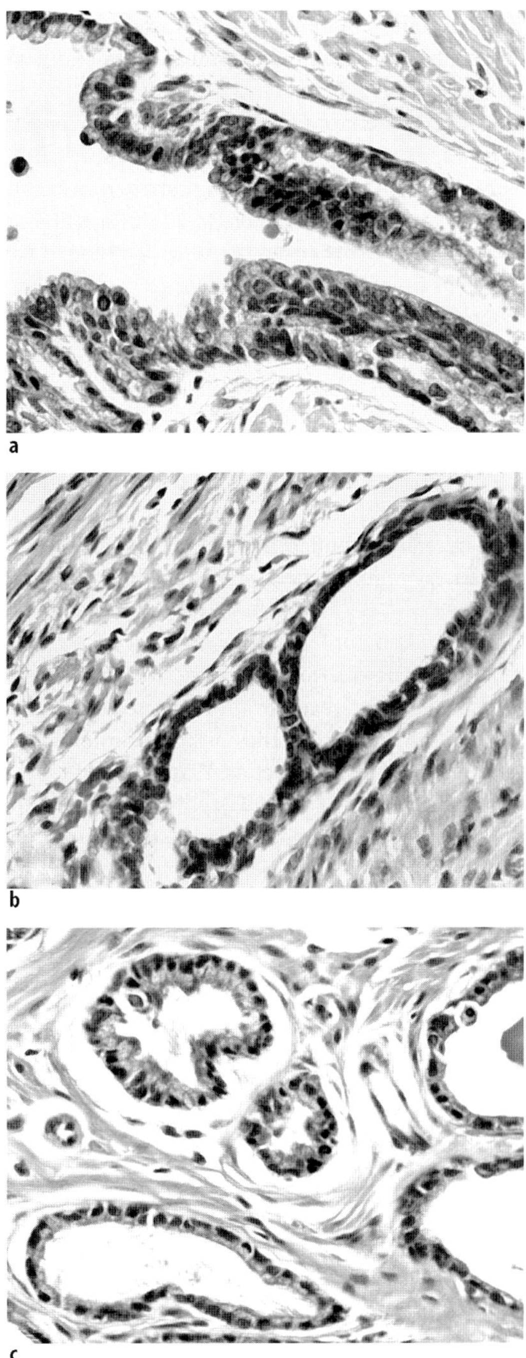

Fig. 19a–c Excretory glandular ducts from proximal (**a**), intermediate (**b**), and distal (**c**) portions. H&E, ×400

Neuroendocrine cells exhibit abundant differences in quantity. Their distribution is irregular and more evident in ducts than in acini (Rodríguez Ramos 2001; Rodriguez et al. 2003; Ingelmo 2005).

The embryologic development of prostatic NECs is controversial, and there are two possible hypotheses. According to the theory of neurogenic origin (Aumuller et al. 1999), the neuroendocrine population originates from the neuro-ectodermal crest, migrating during the embryonic stage to the urogenital sinus (Pearse and Polak 1971). Other authors have proposed the stem cell theory, which gives a common origin to all the cells included in the prostate epithelium: basal, intermediate, luminal, and neuroendocrine (Bonkhoff and Remberger 1996; Schalken and van Leenders 2003).

As in human prostate, two types of neuroendocrine cells can be identified in rat prostate, open and closed. Both possess cytoplasmic processes (dendrite like) that are in contact with epithelial cells or with other NECs. The closed cells do not make contact with the lumen, but the cytoplasmic processes of the open cells can eventually reach the acinar lumen, showing long stereocilia at their apical zones. These microvilli can serve as sensors for the luminal contents and send signals to regulate prostatic secretion (Montuenga et al. 2003; Marraco 2004).

The distribution of dendritic processes in contact with the neighboring epithelial cells suggests paracrine regulation. Likewise, the mutual contact among the processes and the expression of receptors for calcitonin in groups of neuroendocrine cells suggests both paracrine and autocrine interactions. Although the NECs express calcitonin receptors (Wu et al. 1996), they do not show expression of androgen receptors (Bonkhoff et al. 1993); this can be explained because castration in rodents (Guinea pig) does not affect the population of neuroendocrine cells (Acosta et al. 2001). Likewise, although they do not express the antiapoptotic factor Bcl-2 (Xue et al. 1997), they express the antiapoptotic protein survivin (Ambrosini et al. 1997); this supports the hypothesis that neuroendocrine cells are impervious to apoptosis and can bear stress-producing conditions (i.e., castration, denervation) (Xing et al. 2001).

Ultrastructural analysis reveals a wide diversity of neurosecretory granules that suggest a variety of functionally different subtypes. The content of the granules includes serotonin, histamine, chromogranins, somatostatin, cholecystokinin, NPY, proadrenomedullin, CGRP, and bombesin, among others. These molecules can regulate the growth, differentiation, and homeostasis of prostate tissue through paracrine, autocrine, endocrine, exocrine, and neurocrine mechanisms. The presence of neuropeptides in the luminal contents would indicate exocrine secretion (di Sant'Agnese and Mesy Jensen 1984a).

The NECs are distributed either as isolated cells or in little groups among the epithelial cells of acini or ducts, and can establish desmosome-like junctions with the epithelium. At present, it is supposed that the neuroendocrine cells could represent an intermediate link between the autonomic innervation of the prostate and the epithelial cells. Structural studies in human prostate have demonstrated

NECs closely associated with neuroendings, suggesting a direct nervous connection (Abrahamsson 1999; Acosta et al. 2001).

6.1.1
Neuroendocrine Cells in the Acini of the Rat Prostate

The presence of neuroendocrine cells in the acini is under discussion. The first study on the presence of acinar NECs (Angelsen et al. 1997), analyzed the possible existence of these cells in dog, cat, rat, and guinea pig prostate. These authors only found NECs in guinea pigs. A more recent study on this species (Acosta et al. 2001), detected cells positive for serotonin and chromogranin A, located in the acini from ventral and dorsal lobes.

Neuroendocrine cells immunoreactive to chromogranin were also detected in dorso-lateral acini of rat prostate after treatment with testosterone, which provoked proliferative epithelial lesions. Other authors (Jimenez et al. 1999) found few cells resembling the typical neuroendocrine cell type, immunoreactive to adrenomedullin, proadrenomedullin, and serotonin in the ventral acini.

6.1.2
Neuroendocrine Cells in the Ductal System of the Rat Prostate

Ductal neuroendocrine cells expressing adrenomedullin, proadrenomedullin, serotonin, and chromogranin (Jimenez et al. 1999; Rodriguez et al. 2003) were shown intermingled with the epithelial cells lining the ducts from all prostate lobes; the NECs seem to be more abundant in the portions of the ducts proximal to the urethra.

In contrast to the situation observed in humans (Rode et al. 1985; Martin et al. 2000), no PGP 9.5-immunoreactive neuroendocrine cells were found in rats by some authors (Rodriguez et al. 2003); however, others (Ingelmo 2005) described occasional PGP-immunoreactive neuroendocrine cells in rat prostate. Conversely to the situation observed in human prostate, there was no evidence about the presence in rat of NECs immunoreactive to somatostatin, arginine vasopressin, synaptophysin, CGRP, etc.; in this respect, some authors (Uchida et al. 2005) describe the characterization in vitro of a prostatic neuroendocrine cell line established from a neuroendocrine carcinoma of a transgenic mouse that produced serotonin and somatostatin in conditioned medium. There is little information about the potential role of different types of prostate NECs with different neurotransmitters. For example, bombesin might be involved in the growth and differentiation of glandular epithelium (di Sant'Agnese 1986), parathyroid hormone might be involved in modulation of Ca^{2+} concentration in prostate secretions (Iwamura et al. 1994c) and also in growth processes (di Sant'Agnese 1986), CGRP could play a regulating role in the contractility of prostate smooth muscle (di Sant'Agnese et al. 1989), and serotonin might stimulate the proliferation of prostate stromal cells (Cockett et al. 1993).

Serotonin is the best marker for visualizing NECs from rat prostate; nevertheless, immunostaining for chromogranin A provides a weak and diffuse signal in the rat,

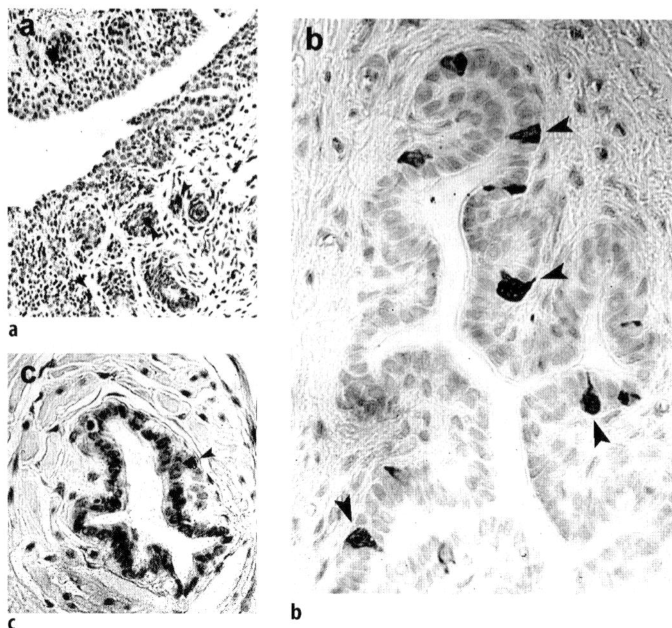

Fig. 20a–c Neuroendocrine cells (*arrowheads*) in the excretory ducts of the rat prostate during postnatal development. **a** Prepubertal rat. **b** Pubertal rat. **c** Aged adult rat. **a** ×100; **b, c** ×400

although it is quite good for demonstrating the presence of neuroendocrine cells in human prostate (Rodriguez et al. 2003). Neuroendocrine cells immunoreactive to serotonin have been detected in prepubertal, pubertal, and adult animals. These cells are located among the columnar cells in the periurethral excretory ducts. No serotonin-immunostained neuroendocrine cells were observed in the acinar epithelium throughout the prostate lobes. Neuroendocrine cells appear to be poorly distributed in prepubertal rats but were more abundant in both pubertal and young adult animals (Fig. 20a,b), and again infrequent in older animals (Fig. 20c). They are rounded or triangular, frequently showing apical processes (Fig. 20b).

The morphological patterns observed in neuroendocrine cells immunoreactive to serotonin in rats were similar to those described in humans, but were exclusively present in excretory periurethral ducts. This partially agrees with results obtained by other authors (Jimenez et al. 1999), who also found some neuroendocrine cells in ventral acini. In this respect, some authors also observed neuroendocrine cells in rat prostate acini only after treatment with testosterone (Angelsen et al. 1999).

6.1.3
Quantitative Changes of Neuroendocrine Cells during Postnatal Development

Stereological methods are very appropriate to quantify both relative (numerical densities) or absolute (number of cells per organ) numbers of neuroendocrine cells. These techniques ensure an unbiased and assumption-free estimate of cell magnitudes (Mayhew and Gundersen 1996; Howard and Reed 2005a).

The N_V (numerical density) of serotonin-immunoreactive neuroendocrine cells was significantly increased (P<0.05) in pubertal animals compared with prepubertal rats. Moreover, this parameter is also significantly higher in pubertal rats than in both young and adult animals. The N_V of serotonin-immunoreactive neuroendocrine cells in aged animals decreases to a level similar to that observed in prepubertal rats (Fig. 21a).

The absolute number of neuroendocrine cells immunostained for serotonin increases significantly from prepubertal to young adults, while this parameter decreases in aged adults. The absolute number of serotonin-immunoreactive cells was significantly higher in young adult rats than in the other groups (Fig. 21b).

The predominance of neuroendocrine cells in rat periurethral ducts agrees with the findings of other authors (Aumuller et al. 2001) in human prostate. These authors indicate the presence of a density gradient of neuroendocrine cells, with the highest density in the large collicular ducts and almost no cells in subcapsular peripheral acini.

Quantitative analysis revealed an increase in the relative number of neuroendocrine cells immunoreactive to serotonin from prepubertal to pubertal stages, and the N_V of serotonin-immunostained cells then experiences a progressive decrease during adulthood: Aged rats showed values similar to those observed in prepubertal animals. However, the largest absolute number of neuroendocrine cells immunostained for serotonin was observed in young adults. One explanation for this may be that the prostate in these animals shows a significant increase in volume in comparison with that prepubertal rats. It is interesting to note that the size of the prostate gland does not decrease in aged rats, but the absolute number of serotonin-immunoreactive cells in those animals drops to a level similar to that found in the pubertal group. This suggests an actual diminution in the neuroendocrine cell population with aging, and this event might be related to the decrease in androgenic levels, suggesting some androgen dependence in maintaining the serotoninergic neuroendocrine cell population. This finding does not agree with the observations of other authors (di Sant'Agnese et al. 1987) in guinea pig prostate, where an increase in the number of neuroendocrine serotoninergic cells per unit area was detected in aged animals. These differences could be attributed to species-specific differences and also to the different methodology employed in quantification, because the evaluation of cell number per unit area does not express the number of cells per prostate and is somewhat flawed from a stereologic point of view (Howard and Reed 2005a).

There is a parallelism between prostate gland maturation and evolution of N_V of serotoninergic neuroendocrine cells in rats, because secretory activity starts at 12 days postpartum and rises to adult levels in pubertal animals (Aumuller 1991; Hayashi et al. 1991). This also supports an androgenic effect on the development of neuroendocrine cells. However, there are observations (Acosta et al. 2001) suggesting that neuroendocrine cells from guinea pig prostate seem to be independent of innervation and androgens.

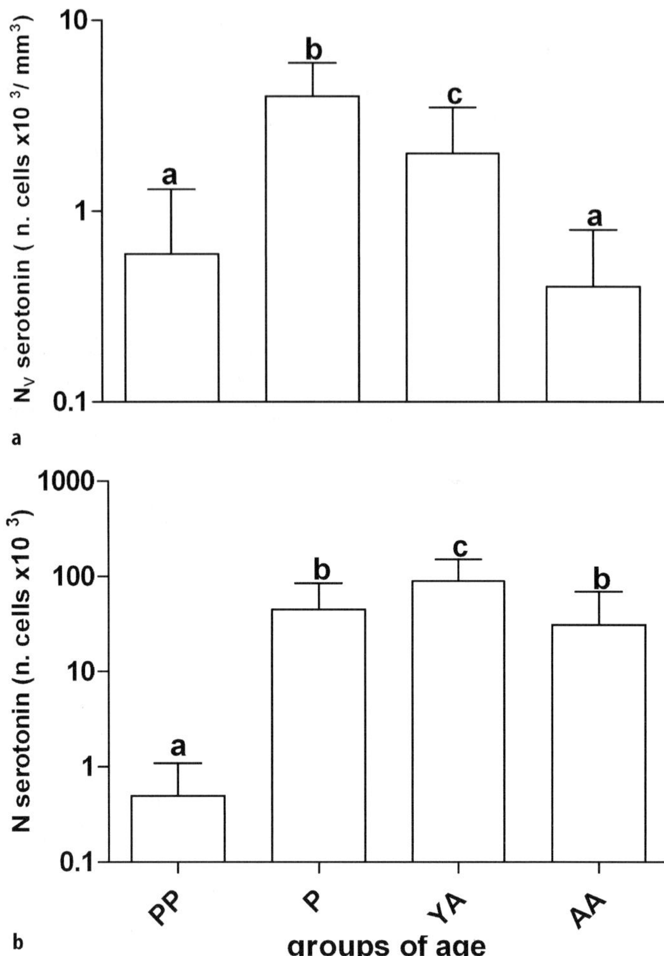

Fig. 21a, b Bar diagrams indicating means ± SD. **a** Numerical density (N_V) of neuroendocrine cells immunostained by serotonin, expressed in number of immunoreactive cells x 10^3/mm^3 of epithelial tissue from prepubertal (*PP*), pubertal (*P*), young adult (*YA*), and aged adult (*AA*) rats. **b** Absolute number of neuroendocrine cells immunostained to serotonin expressed in number of immunoreactive cells per 10^3 from prepubertal (*PP*), pubertal (*P*), young adult (*YA*), and aged adult (*AA*) rats. In each graph, the *letters* over each *bar* indicate the significance: *Bars* labeled by different *letters* show significant differences ($P<0.05$)

These results differ from those described in human prostate by some authors (Cohen et al. 1993): The number of neuroendocrine cells found in periurethral ducts from human prostate increases progressively during adulthood and does not seem to be androgen dependent. On the other hand, the neuroendocrine cells found in the acini might be androgen dependent, appearing during postpubertal stages, increasing until the sixth decade of life, and then decreasing with aging.

There are recent studies (Xue et al. 2000; Aumuller et al. 2001) on distribution and numerical density of neuroendocrine cells in human prostate at fetal, prepubertal, and young adult stages: The neuroendocrine cells appear at 13 weeks of gestational age in both acini and excretory ducts. There is no variation in their numerical density from fetal to young adult stages, and in contrast to the situation observed in rats, there do not seem to be any relevant changes in relation to sexual maturity.

The exclusive presence of the neuroendocrine cells among the columnar cells from periurethral ducts might be related to a possible role of the neuroendocrine cells in regulating the excretion of prostatic fluid toward the urethra. In this sense, serotonin might be able to stimulate the contraction of the muscular layers of periurethral ducts in a way similar to that observed in the intestinal wall, where serotonin stimulates the peristalsis modulating cholinergic activation and noradrenergic inhibition (Berezina 1998). Moreover, it has been demonstrated that serotonin and agonists of serotoninergic receptors are able to induce epididymal and prostate contraction in rats (Minker and Bartha 1981; Killam et al. 1995). There was also an interesting parallelism between the distribution of rat NECs (in the periurethral region) and the predominant abundance of human NECs in the transition zone (Santamaria et al. 2002). Although the origin and functional significance of rat and human NEC locations might be different, at least there was a topological similarity between human transition zone and rat periurethral ducts.

Some relevant conclusions can be drawn from these observations: (a) The immunoreactivity and distribution of neuroendocrine cells differ in rat prostate in comparison to human prostate; they are weakly immunopositive to chromogranin A, do not immunostain to PGP 9.5, and only express serotonin immunoreactivity. They are exclusively found among columnar cells from periurethral ducts and do not appear in acinar epithelium. (b) The changes in the numbers (relative and per prostate) of rat neuroendocrine cells during postnatal development may be influenced by androgens. (c) Their localization in periurethral ducts could be related to the regulation of progress of prostate secretions to the urethral lumen.

6.2
Hormonal Influences on Neuroendocrine Cells in Rat Prostate

Androgens are required for the development and maintenance of rat prostate (George and Peterson 1988). It is well known that pharmacologic castration induces relevant morpho-functional changes in both epithelial and mesenchymal compartments of rat prostate; nevertheless, the possible effect of pharmacological blocking of androgen receptors on changes in NECs population and peptidergic innervation is not well established. On the other hand, the action of nonsteroid hormones on prostate is a potential field of interest. For example, a relationship between levels of prolactin and the increase of prostate pathology in humans has been described (Bartke 2004); moreover, prolactin stimulates the androgen-independent growth of rat prostate cells in vitro (Ahonen et al. 1999).

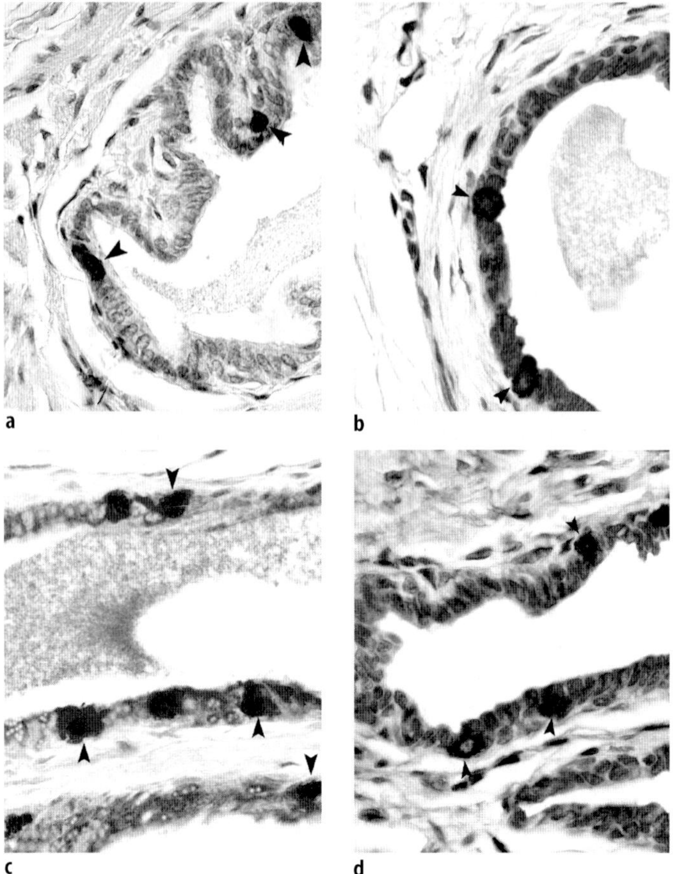

Fig. 22a–d Neuroendocrine cells immunostained by serotonin, observed in the glandular prostate ducts (*arrowheads*), from an untreated rat (**a**), a pharmacologically castrated rat (cyproterone acetate exposed) (**b**), a prolactin-treated animal (**c**), and a rat exposed to prolactin plus cyproterone acetate (**d**). ×400

Recent studies (Ingelmo 2005) have indicated that pharmacologic castration with cyproterone acetate significantly decreases the absolute number of neuroendocrine cells immunoreactive to serotonin in the rat prostate (Figs. 22a–c and 23). These results disagree with the presumptive androgen independence of neuroendocrine cells reported by other authors in both rodents and humans (Bonkhoff et al. 1993; Acosta et al. 2001).

Prolactin administration to rats, either intact or pharmacologically castrated, induces a significant increase in the number of NECs immunostained for serotonin (Fig. 23). This agrees with the presence of prolactin receptors reported in the rat prostate (Nevalainen et al. 1996); thus the number of neuroendocrine cells in rat prostate can be regulated by prolactin levels in an androgen-independent way.

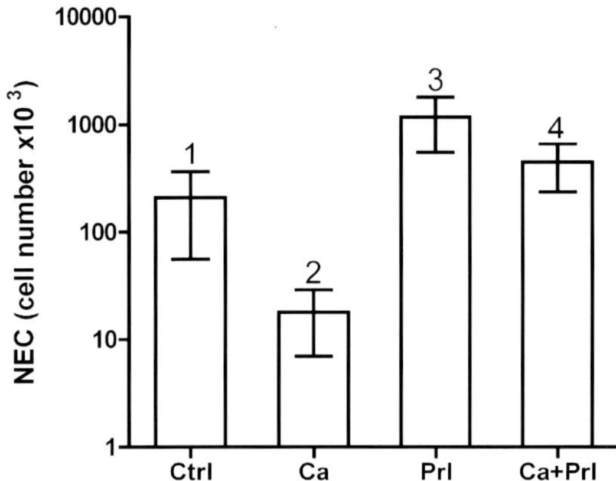

Fig. 23 Bar diagram indicating means ± SD for absolute number (*NEC*) of neuroendocrine cells immunostained by serotonin expressed in number of immunoreactive cells per 10^3 from untreated rats (*Ctrl*), pharmacologically castrated rats (cyproterone acetate exposed) (*Ca*), prolactin-treated animals (*Prl*), and rats exposed to prolactin plus cyproterone acetate (*Ca+Prl*). The *numbers* over each *bar* indicate the significance: *Bars* labeled by different *numbers* show significant differences ($P<0.05$)

7
Innervation of the Rat Prostate
J.M. Pozuelo, R. Rodríguez

7.1
Generalities

The innervation of the rat prostate fundamentally depends on the autonomous or vegetative nervous system and to a minor degree on the somatic voluntary nervous system. The vegetative components are the parasympathetic and orthosympathetic systems (Fig. 24) (Dail 1993; Kepper and Keast 1995; McVary et al. 1998).

The orthosympathetic efferent fibers, proceeding from the intermediolateral horn of the lumbar spinal cord and last thoracic segments (T10–L2), arrive at the paravertebral sympathetic ganglion chain and from here they head for the celiac abdominal plexus from which arise several branches; one of these, the aortic abdominal plexus, including the inferior mesenteric ganglion, heads in the caudal direction to form the hypogastric plexus, which at the level of the promontory bifurcates in the right and left branches to get to the great pelvic plexus or major pelvic ganglions (right and left, respectively) (Purinton et al. 1973; Wanigasekara et al. 2003). Additionally, the pelvic ganglion receives nerve fibers from sacral and coccygeal orthosympathetic chains.

The caudal parasympathetic innervation is provided by the pelvic nerve. The preganglionic parasympathetic fibers have their origin in the superior and medial levels of the sacral cord (S2, S3, and S4) that constitute the pelvic nerve (McVary et al. 1998; Nadelhaft 2003), whose axons head for the major pelvic ganglion.

The pelvic nerve penetrates into the major pelvic ganglion at its dorsal face, while the hypogastric nerve (orthosympathetic) runs next to the ureter and enters into the ganglion through its superior face.

The somato-sensory innervation derives from sacral roots S1–S4, which project part of their fibers at the major pelvic ganglion, accomplishing the somato-sensory innervation of the prostate through the visceral branches of the pudendal plexus, where they join the autonomic innervation to reach the prostate (Pennefather et al. 2000; Acone et al. 2001) (Fig. 24).

The existence of a wide overlap in the innervation of the visceral pelvic organs (urinary bladder, prostate, urethra, and rectum) seems likely at central levels, although not at spinal levels (Orr and Marson 1998; Zermann et al. 2000; Nadelhaft et al. 2002).

Diverse postganglionic branches arise from the major pelvic ganglion, the rectal posterior, the superior urethro-vesical, and the inferior vesico-urethro prostatic branches being the more important. The last branch directs its nerve fibers to the inside of the prostate, together with the prostate blood vessels. The nerve fibers within the prostate are distributed around the blood vessels, in the stroma, surrounding the smooth muscle from ductal walls, and in the vicinity of the acini.

Additionally, the prostate of the rat shows abundant pericapsular and intramural vegetative microganglia (Perez Casas et al. 1985; Sakamoto et al. 1999; Nadelhaft et al. 2002), receiving many postganglionic fibers from the major pelvic ganglion. The intramural microganglia originate from postganglionic fibers that are also distributed within the various prostate compartments.

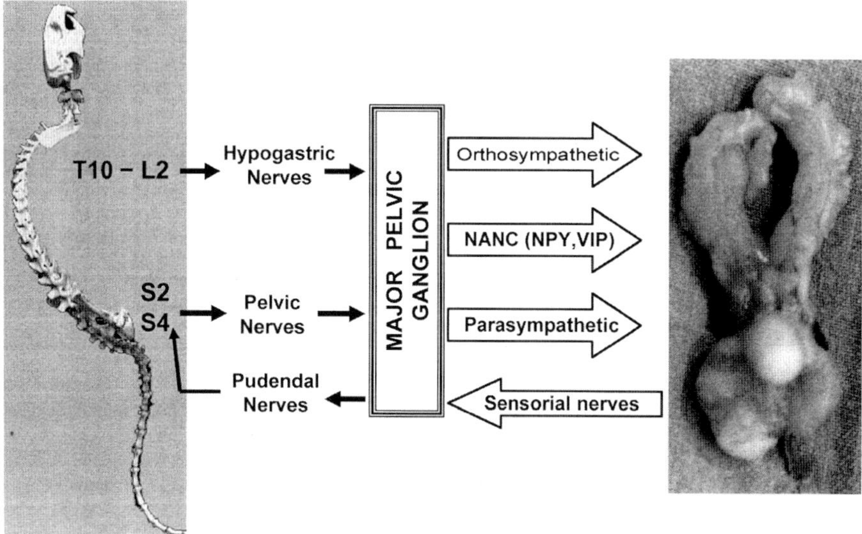

Fig. 24 Scheme showing orthosympathetic, parasympathetic, nonadrenergic, noncholinergic (*NANC*), and sensory innervation in the rat prostate

7.2
Major Pelvic Ganglion

The neuronal composition of the major pelvic ganglion has been widely studied during the last few years (Wanigasekara et al. 2003). The rat pelvic ganglia contain both orthosympathetic and parasympathetic neurons. These two cell groups are present in approximately equal proportions. Approximately 75% of sympathetic neurons are presumed to be noradrenergic, which show tyrosine hydroxylase (TH) immunoreactivity, and they are also immunoreactive for neuropeptide Y (NPY); the remainder contain vasoactive intestinal polypeptide (VIP) immunoreactivity but not TH and they may be cholinergic. Parasympathetic neurons were virtually all non-noradrenergic (TH negative) and were also of two types, with some neurons expressing NPY and others VIP (Keast 1995).

The majority of the projecting neurons are orthosympathetic and contain both noradrenaline and NPY, with smaller nonnoradrenergic populations containing either VIP or NPY. Parasympathetic innervation of rat prostate gland is represented by 20–30% of pelvic ganglion neurons that are nonnoradrenergic (possibly cholinergic). These neurons express both VIP and NPY (Kepper and Keast 1995). Recent studies show a group of small cholinergic neurons immunostained for vesicular acetylcholine transferase (VAChT) and large, mainly adrenergic neurons surrounded by preganglionic cholinergic terminals (Nadelhaft 2003).

Other studies in mice describe the structural and histochemical features of the major pelvic ganglion. Almost all pelvic ganglionic cells are monopolar, and most of them are cholinergic. All contain either NPY or VIP, or both peptides together. Virtually all noncholinergic pelvic ganglion cells are noradrenergic and contain NPY. These studies show that the male mouse pelvic ganglion exhibits some differences from that in the rat, and VIP/NPY colocalization is much more common in the mouse (Wanigasekara et al. 2003).

7.3
The Relevance of the Autonomous Nervous System for Rat Prostate Structure

For many years, prostatic function, structure, maintenance, and growth were considered to be exclusively controlled by the endocrine system. However, a number of works have appeared in the last few years that demonstrate the importance of the autonomous nervous system in the integrity of the rat prostate (Wang et al. 1991a; Martinez-Pineiro et al. 1993).

When bilateral pelvic ganglion denervation was performed in adult rats, a significant decrease in ventral prostate weight, together with histological changes such as an overall reduction in luminal staining, decrease in cell height, and an apparent increase in intracellular vacuoles and intercellular empty spaces, was observed. The effect of denervation on prostate ultrastructure was evidenced by an overall reduction in the number and height of microvilli and a decrease in the height of supranuclear and apical regions of epithelial cells. In summary, results of this study indicate a structural and functional impairment in the chronically denervated rat prostate (Wang et al. 1991b).

Other authors (Martinez-Pineiro et al. 1993) found morphologic and functional changes in the gland epithelium after unilateral denervation of the right pelvic ganglion. The histologic features of the denervated prostate showed an overall decrease in cell height and a reduction in the clear apical area of the supranuclear region. With electron microscopy, the epithelial cells showed a significant decrease in the number of secretory granules, a decrease in the height of the supranuclear region, and fewer and less abundant dilated apical cisternae of endoplasmic reticulum. These changes indicate a modification of secretory activity and might reflect a change in epithelial metabolism.

Unilateral sympathectomy (McVary et al. 1994) leads to decreases in ventral prostate weight and DNA and protein contents in the denervated side, whereas increases in the same parameters were observed in the contralateral side.

7.4
Classic and Peptidergic Innervation

The importance of prostate innervation in normal physiology has been traditionally ascribed to the classic autonomic neurotransmitters noradrenaline and acetylcholine. In the last 15 years, there is evidence suggesting that neuropeptides contained in prostatic autonomic nerves play a role in the regulation of prostate function. The best studied of these neuropeptides is VIP, which is abundant in autonomic rat prostate nerves (Alm et al. 1980; Polak and Bloom 1984; Vega et al. 1990; Properzi et al. 1992; Kepper and Keast 1995; Rodriguez et al. 2005). The presence of high-affinity receptors for VIP together with the occurrence of VIP-containing neurons innervating the prostate, the finding of a VIP-stimulated cAMP system in rat prostatic epithelial cells (Carmena and Prieto 1983), and the coexistence of receptors for cholinergic, adrenergic, and peptidergic agents, which can regulate cAMP (Carmena and Prieto 1985), suggests that the functions of prostatic epithelium may be interdependently controlled by multiple neural effectors and that this peptide may be relevant in the physiological regulation of the functions of prostatic epithelium.

The most widely distributed neuropeptide in rat prostatic autonomic nerves is NPY (Properzi et al. 1992; Kepper and Keast 1995; Rodriguez et al. 2005). The function of NPY in rat prostate is not sufficiently known. Some studies on the effects of NPY in the nervous system have demonstrated a role for NPY in inhibiting the effects of other neurotransmitters. NPY has been shown to inhibit the release of noradrenaline in rat and guinea pig nerve terminals from vas deferens (Torres et al. 1992). NPY also acts by impeding the effects of several neurotransmitters by inhibiting the activation of adenyl cyclase. This effect is mediated by a receptor-coupled inhibitory G protein. In particular, NPY has been shown to inhibit cAMP production stimulated by alpha or beta adrenergic agonists or by VIP in several cell types such as bovine adrenal chromaffin cells and rat pinealocytes (Harada et al.

1992; Zhu et al. 1992). Moreover, NPY inhibits VIP-stimulated cAMP accumulation and VIP-stimulated adenyl cyclase activity in isolated rat ventral prostatic epithelial cells through a guanine nucleotide regulatory G protein (Solano et al. 1994).

These findings suggest that NPY in the prostate could act by modulating the effects of VIP or other neurotransmitters on epithelial cells, but the effects of NPY on prostate secretion have not been extensively studied.

7.4.1
Intraprostatic Innervation

As early as 1979, it was found that the rat ventral prostate receives dual auto-nomic innervation. Adrenergic fibers, which form the majority of the nerves, were seen in close contact with the smooth muscle cells around both the prostatic acini and secretory ducts. The nonadrenergic fibers (probably cholinergic), which were fewer in number, did not experience such intimate contacts with the muscle cells and did not establish synapsis with epithelial cells (Vaalasti and Hervonen 1979).

Other authors (Kepper and Keast 1995) found a dense plexus of varicose axons stained for either TH (a marker for noradrenergic fibers) or NPY associated with the acinar epithelium. Many axons contained both TH and NPY, but NPY axons consistently predominated. VIP axons were associated with most acini but formed a more scanty plexus than the NPY or TH axons.

Other histochemical studies carried out on rat and guinea pig prostate revealed the presence of acetylcholine and noradrenaline fibers in the fibromuscular stroma. On the other hand, only acetylcholine fibers were seen innervating the epithelium (Lau et al. 1998).

Recent studies performed on rats show acetylcholine fibers (vesicular acetyl-choline transferase marker) in the inner epithelial layer and some muscarinic subtype receptors in the outer muscle layer surrounding the prostatic acini. Adrenergic fibers (vesicular monoamine transporter marker) have also been ob-served in the inner secretory layer and outer muscle layer (Wang et al. 1991a; Nadelhaft 2003).

The behavior of both cholinergic (nicotinic and muscarinic) and adrenergic (al-pha and beta) receptors in rat prostate has also raised interest from physiologic and pharmacologic points of view. In this sense, a group of authors made a quantitative analysis of rat prostate secretion in response to hormonal and pharmacological manipulation, concluding that α_1-adrenergic stimulation causes secretion by con-traction of prostatic smooth muscle, whereas cholinergic stimulation causes a low but maintained secretory effect, which appears to be due to a direct stimulation of epithelial secretion (Wang et al. 1991b). It seems that the neurotransmission for the prostate smooth muscle in the three studied species, rat, guinea pig, and mouse, is predominantly sympathetic and noradrenergic (Lau et al. 1998; Wanigasekara et al. 2003), but there is some participation of acetylcholine, acting on muscarinic receptors (Lau and Pennefather 1998; Lau et al. 1998).

A more recent histological and pharmacological study that investigate the innervation of the prostate in a number of species (including rat) demonstrated the presence of noradrenergic innervation in the stroma but not in secretory acini. Stimulation of these nerves caused contractions of prostate smooth muscle that was inhibited by α_1-adrenoceptor antagonists. The acetylcholine-positive nerves were seen in both stroma and epithelium. Some muscarinic receptors were implicated in prostate contraction. The presence of VIP, NPY, and other neuropeptides in the prostatic stroma indicate they may act as cotransmitters or modulators from autonomic effector nerves (Pennefather et al. 2000).

A study found muscarinic receptors in the outer muscle layer of the acini. All these findings provide new evidence of a main cholinergic influence over the rat prostate (Nadelhaft 2003).

Summarizing, there are abundant studies describing the distribution of rat prostate nerve fibers. However, there are few works that explore a quantitative approach to the peptidergic innervation of rodent prostate using stereological unbiased tools (Chow et al. 1997; Rodriguez et al. 2005).

7.5
Postnatal Evolution of the Peptidergic Innervation of the Rat Prostate

The presence, distribution, and quantification of nervous fibers has been studied with a combination of immunohistochemistry and stereological unbiased methods for measuring of nerve fiber length (Rodriguez et al. 2005). This study was performed throughout all the regions of the rat prostate during postnatal development, using protein gene product 9.5 (PGP 9.5) as general marker for nervous tissue and NPY and VIP as functional innervation markers.

7.5.1
Immunohistochemical Findings

Nerves immunoreactive to PGP 9.5 were observed for each age group, in each prostate region, including postganglionic neurons from ganglia associated to periprostatic capsule (Fig. 25a) and periglandular or interglandular compartments (Fig. 25b).

NPY-immunoreactive nerves were observed from the postnatal period in each prostate region and tissular compartment. NPY fibers from the periductal compartment were more notable in both pubertal and adult rats (Fig. 25c).

Immunoreactivity to VIP was already detected in periglandular and interglandular nerves in prepubertal rats and also in clusters of periprostatic neurons and in nerve bundles from prepubertal and pubertal animals (Fig. 25d,e). VIP immunostaining in periglandular fibers was evident throughout postpubertal development (Fig. 25f). Neuronal, fascicular, and periglandular immunoreactivity to VIP was remarkable in aging animals (Fig. 25g).

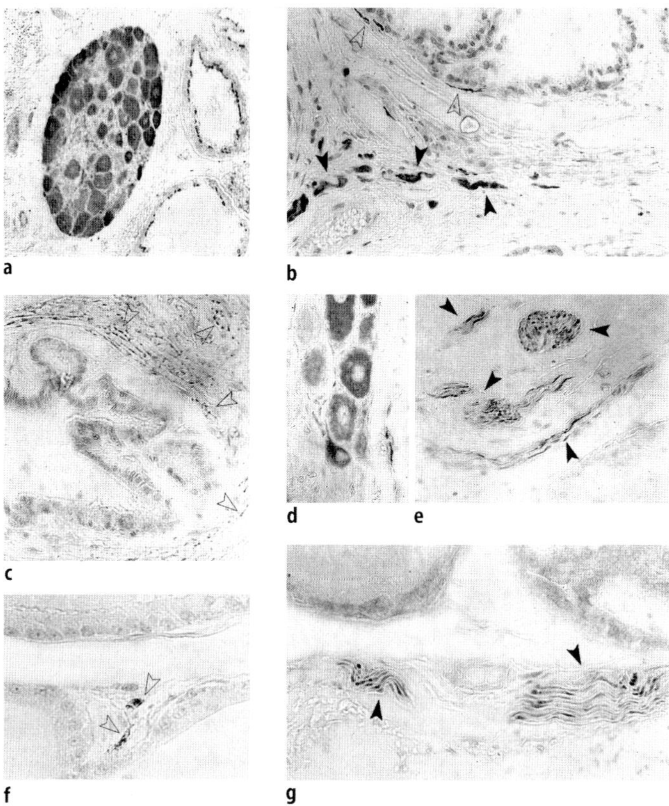

Fig. 25a–g PGP 9.5-, NPY-, and VIP-immunoreactive fibers in the different regions of the rat prostate during postnatal development. **a** PGP-immunostained neurons in a periprostatic ganglion from a prepubertal animal. **b** Periglandular (*black arrowheads*) and interglandular (*empty arrowheads*) PGP 9.5-immunoreactive fibers from a pubertal rat. **c** Periductal fibers immunoreactive to NPY (*empty arrowheads*) in pubertal animals. **d** VIP-immunostained neurons in a periprostatic ganglion from a pubertal rat. **e** Nerve bundles immunoreactive to VIP (*arrowheads*) in the interglandular stroma of a prepubertal animal. **f** Periglandular fibers (*empty arrowheads*) immunostained to VIP in a young adult. **g** Nerve bundles (*arrowheads*) immunoreactive to VIP in the periglandular stroma of an aged adult. ×400

7.5.2
Quantitative Findings

The length density of PGP 9.5-immunoreactive fibers (i.e., length of nerve fibers per unit of volume) remains unchanged during postnatal development (Fig. 26a). Therefore, the relative number of nerve fibers, independent of the type of innervation, was constant in the prostate throughout the postnatal life of the rat. However, within this global population of nerve fibers, changes in the density of fibers immunoreactive to some neuropeptides (VIP, NPY) in relation to the pubertal outcome have been observed (Fig. 26b,c). These changes were evident at the level

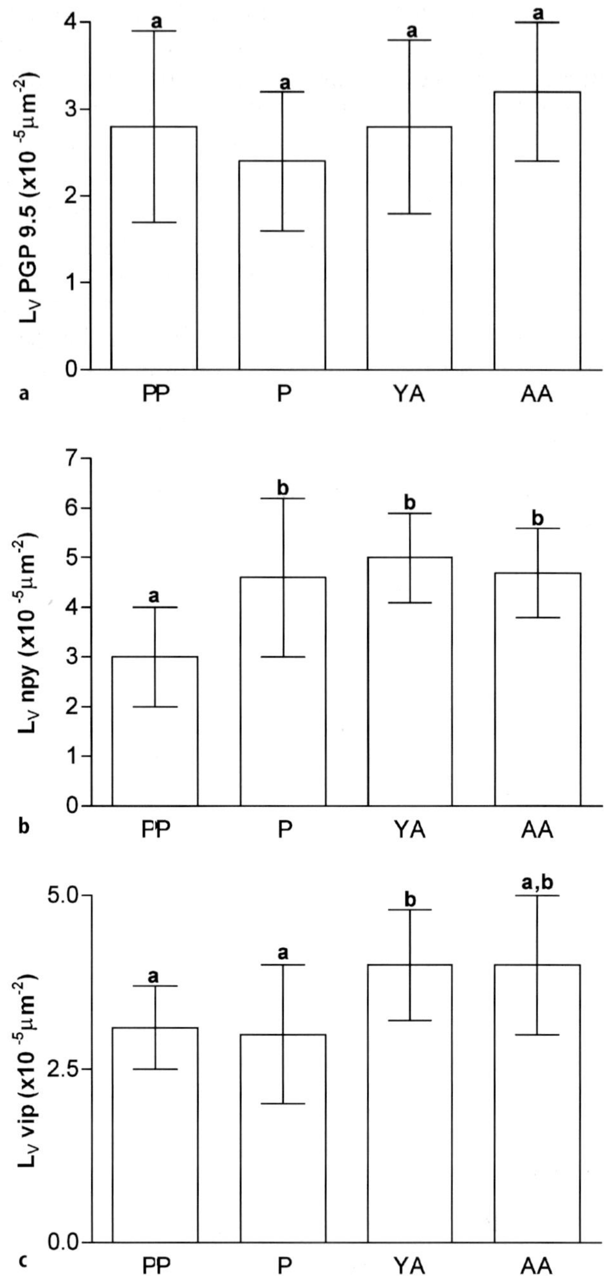

Fig. 26a–c Bar diagrams indicating means ± SD. Length density (L_V) of nerve fibers immunostained by PGP 9.5 (**a**), NPY (**b**), and VIP (**c**) expressed in length ($\times 10^{-5}$ µm) of immunoreactive fiber per µm^3 of stromal volume from prepubertal (*PP*), pubertal (*P*), young adult (*YA*), and aged adult (*AA*) rats. In each graph, the *letters* over each *bar* indicate the significance: *Bars* labeled by different *letters* show significant differences ($P<0.05$)

of excretory ducts and consisted of an increase of NPY- and VIP-immunostained fibers. It seems that changes in density of VIP and NPY fibers are mostly related to increased synthesis. These neuropeptides probably modulate the androgen action on prostate epithelial cells (Gkonos et al. 1995b); therefore, its synthesis probably changes in relation to androgenic function during prostate development. These findings agree with other studies in the urogenital tract from rats (Properzi et al. 1992) or in sexual accessory glands from hamsters (Chow et al. 1997).

The abundance of peptidergic innervation around the excretory ducts is related to the modulation of contractility of the ductal wall (Pennefather et al. 2000; Ventura et al. 2002) necessary for the excretion of prostate fluid during ejaculation (Iwata et al. 2001). In this sense, the particular abundance of innervation in prostate ducts might be associated with the exclusive presence of serotoninergic neuroendocrine

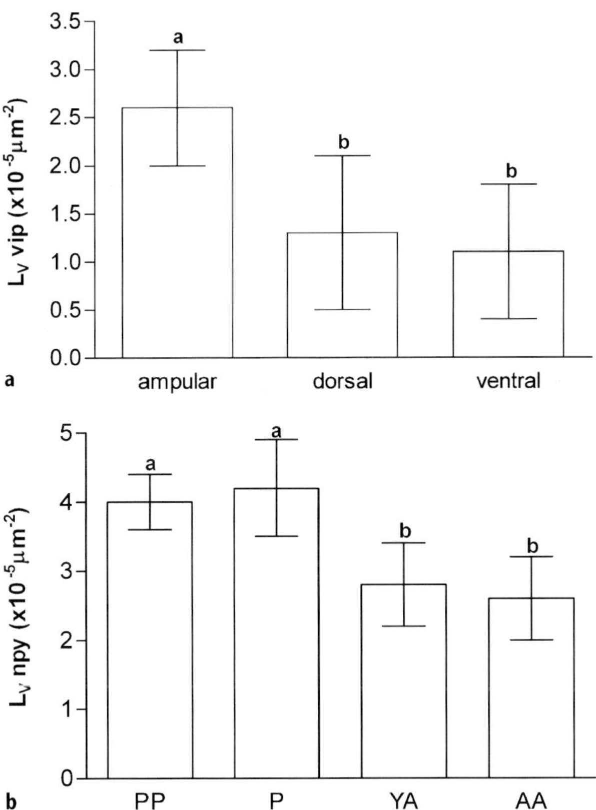

Fig. 27a, b Bar diagrams indicating means ± SD. Length density (L_V) of nerve fibers, expressed in length (×10^{-5} µm) of immunoreactive fiber per µm^3 of periglandular stromal volume. **a** VIP periglandular fibers from ampular, dorsal, and ventral regions in pubertal rats. **b** NPY periglandular fibers in prepubertal (*PP*), pubertal (*P*), young adult (*YA*), and aged adult (*AA*) rats. In each graph, the *letters* over each *bar* indicate the significance: *Bars* labeled by different *letters* show significant differences (*P*<0.05)

cells among the columnar cells from periurethral ducts and their possible role in regulating the excretion of prostatic fluid toward the urethra (Rodriguez et al. 2005).

The parallelism between the increase of neuroendocrine cells and both NPY- and VIP-immunoreactive fibers, after or around puberty, suggesting an androgenic effect on development of prostate innervation, is remarkable (Rodriguez et al. 2003). The periglandular compartment of the ampular gland was the zone most densely populated by VIP-immunoreactive nerve fibers (Fig. 27a). The presence of these subepithelial nerves in the ampular acini suggests a role for VIP in the regulation of their secretory activity (Juarranz et al. 2001; Ventura et al. 2002). This agrees with the detection of VIPergic receptors in the epithelium of rat prostate (Carmena and Prieto 1985; Carmena et al. 1986a,b, 1988).

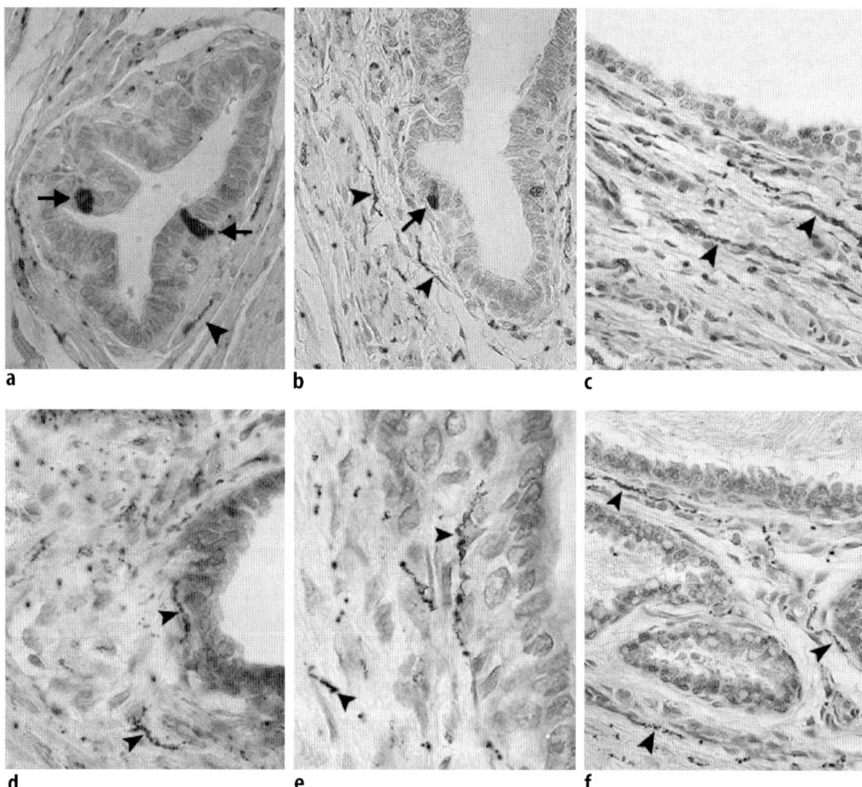

Fig. 28a–f PGP 9.5-immunoreactive fibers around the wall of the excretory ducts (*arrowheads*) from an untreated rat (**a**), a pharmacologically castrated rat (cyproterone acetate exposed) (**b**), and a prolactin-treated animal (**c**). Occasional neuroendocrine cells immunoreactive to PGP were also seen in **a** and **b** (*arrows*). Periductal NPY immunoreactive nerve fibers are seen in a prolactin-treated animal (**d**). Periductal VIP immunoreactive nerve fibers are seen in a cyproterone acetate-treated animal (**e**) and in a prolactin-treated rat (**f**). ×400

It was interesting to note the decrease of L_V NPY in the periglandular compartment after puberty (Fig. 27b). Age-dependent reduction of prostatic autonomic innervation was indicated almost 50 years ago (Casas 1958), and age-related diminution in peripheral autonomic innervation has previously been described in several organs; that is, the aging rat has a reduced sympathetic supply to the urinary tract (Warburton and Santer 1994). One possible cause of such a reduction is the depletion of detectable neuropeptides, because neurons may become less active (Cowen 1993). Other authors (Properzi et al. 1992) have also observed an increase of NPY-immunoreactive fibers in pubertal rat prostate, but these findings were maintained during adult life. This discordance with the present results might be attributed to an absence of rigorous quantification, because other authors (Chow et al. 1997), using stereological unbiased methods for evaluation of NPY innervation in the hamster prostate, agree with the results of this study.

The following conclusions can be made: (a) The relative number of global nerve fibers in rat prostate, detected by PGP 9.5, does not change during postnatal development. Nevertheless, there were significant changes in the NPY and VIP subpopulations of nerve fibers, revealing their increase in periurethral ducts in the pubertal stage. (b) The abundance of peptidergic innervation around the excretory ducts might be related to the modulation of contractility of the ductal wall necessary for the excretion of prostate secretions during ejaculation. (c) The periglandular compartment from the ampular prostate was the most densely innervated in comparison with dorsal and ventral prostate, which might be related to the abundance of smooth muscle in the ampular region. (d) The parallelism between the increase of neuroendocrine cells and both NPY- and VIP-immunoreactive fibers, after or around puberty, suggests an androgenic effect on the development of innervation of periurethral ducts in rat prostate.

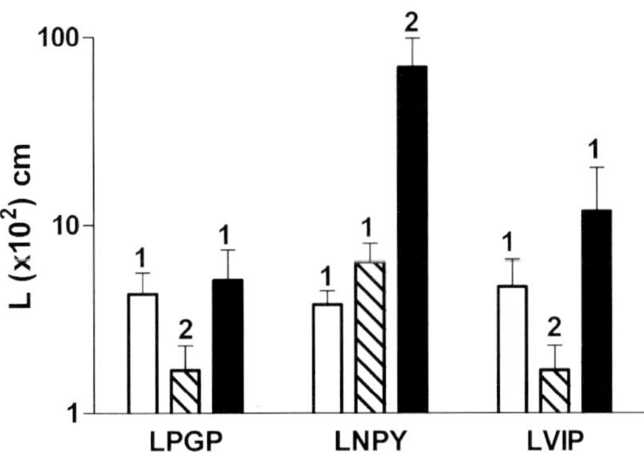

Fig. 29 Bar diagram indicating means ± SD of absolute length (*L*) expressed in centimeters per 10^2 nerve fibers immunostained by PGP 9.5 (*LPGP*), NPY (*LNPY*), and VIP (*LVIP*) from untreated rats (*empty bars*), pharmacologically castrated rats (cyproterone acetate exposed) (*striped bars*), and prolactin-treated animals (*black bars*). The *numbers* over each *bar* indicate the significance: *Bars* labeled by different *numbers* show significant differences (*P*<0.05)

7.6
Hormone Influence and Peptidergic Innervation

Pharmacologic blocking of androgen receptors by cyproterone acetate (Ingelmo 2005) causes a significant decrease of the absolute length of nerve fibers immunostained for PGP 9.5; this was mainly due to a decrease of the length of VIP fibers (Figs. 28a–f and 29). There are a number of mechanisms that explain how the blocking of androgenic stimulus influences the VIPergic innervation. For example, it is known that in rat prostate, the receptor-effector VIP system is enhanced by androgen presence (Juarranz et al. 1994). Conversely, the epithelial androgenic receptors modulate the expression of the VIP receptors (Gkonos et al. 1995b).

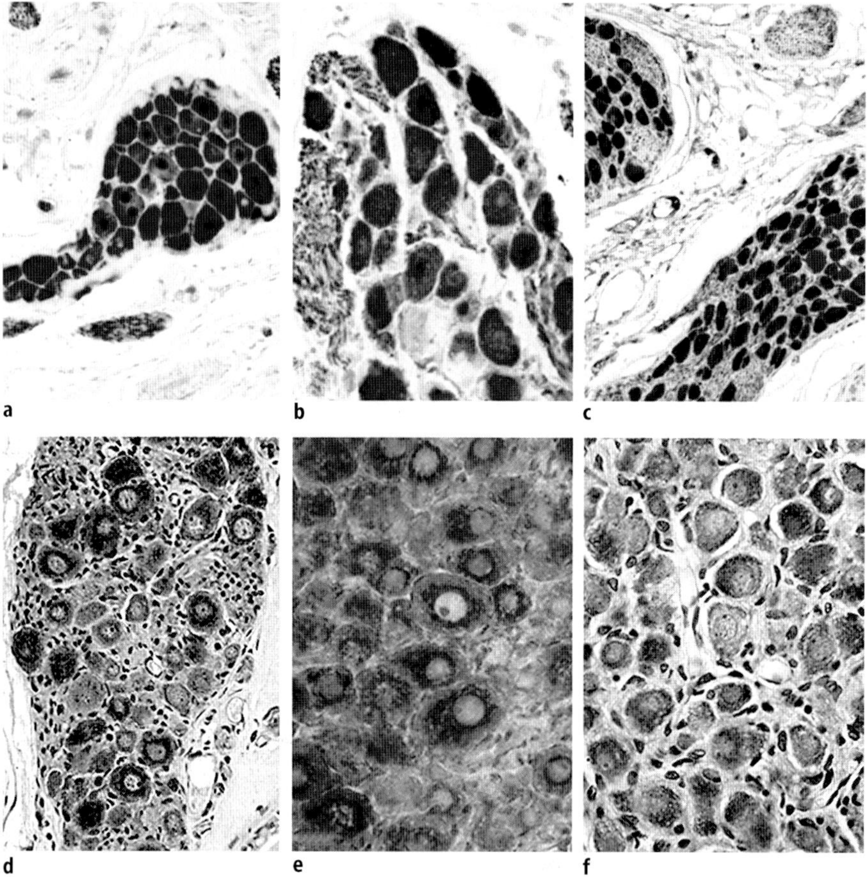

Fig. 30a–f Intraprostatic ganglionic neurons immunostained by PGP 9.5 from untreated rats (**a**), cyproterone acetate-exposed rats (**b**), and prolactin treated-animals (**c**) (×120). NPY-immunoreactive intraprostatic ganglionic neurons from untreated rats (**d**) (×140) and prolactin-treated animals (**e**) (×240). Ganglionic neurons immunoreactive to VIP from a prolactin-exposed rat (**f**) (×200)

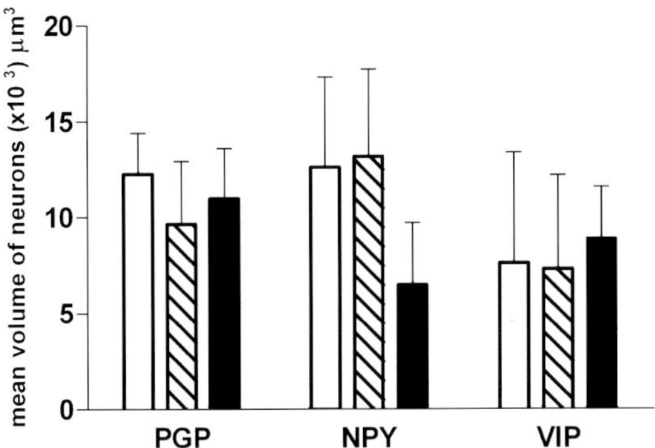

Fig. 31 Bar diagram indicating means ± SD of mean volume of perikaryon, expressed in μm^3 per 10^3, from ganglionic neurons immunostained by PGP 9.5 (*PGP*), NPY (*NPY*), and VIP (*VIP*) from untreated rats (*empty bars*), pharmacologically castrated rats (cyproterone acetate exposed) (*striped bars*), and prolactin-treated animals (*black bars*). There are no significant differences among treatments

However, treatment with prolactin did not change the number of PGP 9.5-positive nerve fibers , although the trophic action of prolactin might selectively act over some subpopulations of nerve fibers; for example, a increase in the length of NPY fibers was ascertained after administration of prolactin (Fig. 29). Nevertheless, this effect of prolactin cannot compensate for the decrease of nerve fiber size produced by castration (Fig. 29).

The changes produced by androgenic blocking and/or prolactin action do not affect stereologic parameters from intraprostatic ganglia (Fig. 30a–f), such as ganglion size, neuronal number, or perikaryon volume of ganglionic neurons (Fig. 31) Thus it can be ascertained that pharmacologic castration and prolactin exposure exert their effects at the level of intraprostate neuroendings.

References

Abrahamsson PA (1999) Neuroendocrine differentiation in prostatic carcinoma. Prostate 39:135–148

Abrahamsson PA, Dizeyi N, Alm P, di Sant'Agnese PA, Deftos LJ, Aumuller G (2000) Calcitonin and calcitonin gene-related peptide in the human prostate gland. Prostate 44:181–186

Abrahamsson PA, Falkmer S, Falt K, Grimelius L (1989) The course of neuroendocrine differentiation in prostatic carcinomas. An immunohistochemical study testing chromogranin A as an "endocrine marker". Pathol Res Pract 185:373–380

Abrahamsson PA, Wadstrom LB, Alumets J, Falkmer S, Grimelius L (1986) Peptide-hormone- and serotonin-immunoreactive cells in normal and hyperplastic prostate glands. Pathol Res Pract 181:675–683

Acone F, Botti M, Gazza F, Sanna M, Cappai MG, Bo ML (2001) Morphological characteristics and distribution of the autonomic and sensitive innervation of the prostate in some animal species. Ital J Anat Embryol 106:1-11

Acosta S, Dizeyi N, Pierzynowski S, Alm P, Abrahamsson PA (2001) Neuroendocrine cells and nerves in the prostate of the guinea pig: effects of peripheral denervation and castration. Prostate 46:191–199

Adrian TE, Gu J, Allen JM, Tatemoto K, Polak JM, Bloom SR (1984) Neuropeptide Y in the human male genital tract. Life Sci 35:2643–2648

Aguilar E, Fernandez-Fernandez R, Tena-Sempere M, Pinilla L (2004) Effects of peptide YY(3–36) on PRL secretion: pituitary and extra-pituitary actions in the rat. Peptides 25:1147–1152

Aguirre P, Scully RE, Wolfe HJ, DeLellis RA (1984) Endometrial carcinoma with argyrophil cells: a histochemical and immunohistochemical analysis. Hum Pathol 15:210–217

Ahlegren G, Pedersen K, Lundberg S, Aus G, Hugosson J, Abrahamsson P (2000) Neuroendocrine differentiation is not prognostic of failure after radical prostatectomy but correlates with tumor volume. Urology 56:1011–1015

Ahonen TJ, Harkonen PL, Laine J, Rui H, Martikainen PM, Nevalainen MT (1999) Prolactin is a survival factor for androgen-deprived rat dorsal and lateral prostate epithelium in organ culture. Endocrinology 140:5412–5421

Algaba F (1993) Bases morfológicas del desarrollo de la hiperplasia prostática. Patología 26:113–119

Algaba F, Trias I (1995) Neuroendocrine cells in the morphogenesis of prostatic pathology. Arch Esp Urol 48:217–222

Algaba F, Trias I, Lopez L, Rodriguez-Vallejo JM, Gonzalez-Esteban J (1995) Neuroendocrine cells in peripheral prostatic zone: age, prostatic intraepithelial neoplasia and latent cancer-related changes. Eur Urol 27:329–333

Ali M, Johnson IP, Hobson J, Mohammadi B, Khan F (2004) Anatomy of the pelvic plexus and innervation of the prostate gland. Clin Anat 17:123–129

Alm P, Alumets J, Hakanson R, Owman O, Sjoberg NO, Sundler F, Walles B (1980) Origin and distribution of VIP (vasoactive intestinal polypeptide)-nerves in the genito-urinary tract. Cell Tissue Res 205:337–347

Alsat E, Haziza J, Scippo ML, Frankenne F, Evain-Brion D (1993) Increase in epidermal growth factor receptor and its mRNA levels by parathyroid hormone (1–34) and parathyroid hormone-related protein (1–34) during differentiation of human trophoblast cells in culture. J Cell Biochem 53:32–42

Amara SG, Jonas V, Rosenfeld MG, Ong ES, Evans RM (1982) Alternative RNA processing in calcitonin gene expression generates mRNAs encoding different polypeptide products. Nature 298:240–244

Ambrosini G, Adida C, Altieri DC (1997) A novel anti-apoptosis gene, survivin, expressed in cancer and lymphoma. Nat Med 3:917–921

Amorino GP, Parsons SJ (2004) Neuroendocrine cells in prostate cancer. Crit Rev Eukaryot Gene Expr 14:287–300

Angelsen A, Falkmer S, Sandvik AK, Waldum HL (1999) Pre- and postnatal testosterone administration induces proliferative epithelial lesions with neuroendocrine differentiation in the dorsal lobe of the rat prostate. Prostate 40:65–75

Angelsen A, Mecsei R, Sandvik AK, Waldum HL (1997) Neuroendocrine cells in the prostate of the rat, guinea pig, cat, and dog. Prostate 33:18–25

Aprikian AG, Cordon-Cardo C, Fair WR, Reuter VE (1993) Characterization of neuroendocrine differentiation in human benign prostate and prostatic adenocarcinoma. Cancer 71:3952–3965

Arriazu R, Pozuelo JM, Martin R, Rodriguez R, Santamaria L (2005) Quantitative and immunohistochemical evaluation of PCNA, androgen receptors, apoptosis, and glutathione-S-transferase P1 on preoplastic changes induced by cadmium and zinc chloride in the rat ventral prostate. Prostate 63:347–357

Aumuller G (1991) Postnatal development of the prostate. Bull Assoc Anat (Nancy) 75:39–42

Aumuller G, Leonhardt M, Janssen M, Konrad L, Bjartell A, Abrahamsson PA (1999) Neurogenic origin of human prostate endocrine cells. Urology 53:1041–1048

Aumuller G, Leonhardt M, Renneberg H, von Rahden B, Bjartell A, Abrahamsson PA (2001) Semiquantitative morphology of human prostatic development and regional distribution of prostatic neuroendocrine cells. Prostate 46:108–115

Aumuller G, Renneberg H, Leonhardt M, Lilja H, Abrahamsson PA (1999) Localization of protein gene product 9.5 immunoreactivity in derivatives of the human Wolffian duct and in prostate cancer. Prostate 38:261–267

Barnett P (2003) Somatostatin and somatostatin receptor physiology. Endocrine 20:255–264

Barry MJ (1990) Epidemiology and natural history of benign prostatic hyperplasia. Urol Clin North Am 17:495–507

Bartke A (2004) Prolactin in the male: 25 years later. J Androl 25:661–666

Bartsch G, Rohr HP (1980) Comparative light and electron microscopic study of the human, dog and rat prostate. An approach to an experimental model for human benign prostatic hyperplasia (light and electron microscopic analysis)—a review. Urol Int 35:91–104

Benoit G, Merlaud L, Meduri G, Moukarzel M, Quillard J, Ledroux M, Giuliano F, Jardin A (1994) Anatomy of the prostatic nerves. Surg Radiol Anat 16:23–29

Bentel JM, Tilley WD (1996) Androgen receptors in prostate cancer. J Endocrinol 151:1-11

Berezina TP (1998) Effect of serotonin on the contractile activity of the cat duodenum. Ross Fiziol Zh Im I M Sechenova 84:940–948

Berry SJ, Coffey DS, Walsh PC, Ewing LL (1984) The development of human benign prostatic hyperplasia with age. J Urol 132:474–479

Birkhoff JD (1983) Natural history of benign prostatic hypertrophy. In: Hinman F (ed.) Benign prostatic hypertrophy. Springer Verlag. New York. 5–14

Bonkhoff H, Remberger K (1996) Differentiation pathways and histogenetic aspects of normal and abnormal prostatic growth: a stem cell model. Prostate 28:98–106

Bonkhoff H, Stein U, Remberger K (1993) Androgen receptor status in endocrine-paracrine cell types of the normal, hyperplastic, and neoplastic human prostate. Virchows Arch A Pathol Anat Histopathol 423:291–294

Bonkhoff H, Wernert N, Dhom G, Remberger K (1991) Relation of endocrine-paracrine cells to cell proliferation in normal, hyperplastic, and neoplastic human prostate. Prostate. 19:91–98

Brawer MK, Peehl DM, Stamey TA, Bostwick DG (1985) Keratin immunoreactivity in the benign and neoplastic human prostate. Cancer Res 45:3663–3667

Bussolati G, Gugliotta P, Sapino A, Eusebi V, Lloyd RV (1985) Chromogranin-reactive endocrine cells in argyrophilic carcinomas ("carcinoids") and normal tissue of the breast. Am J Pathol 120:186–192

Cabo Tamargo JA, Lopez-Muniz A, Bengoechea GE, Vega Alvarez JA, Perez CA (1985) Microscopic innervation of the prostate (I): Distal vegetative formation. Arch Esp Urol 38:231–241

Carmena MJ, Prieto JC (1983) Cyclic AMP-stimulating effect of vasoactive intestinal peptide in isolated epithelial cells of rat ventral prostate. Biochim Biophys Acta 763:414–418

Carmena MJ, Prieto JC (1985) Cyclic AMP response to vasoactive intestinal peptide and beta-adrenergic or cholinergic agonists in isolated epithelial cells of rat ventral prostate. Biosci Rep 5:791–797

Carmena MJ, Recio MN, Prieto JC (1988) Influence of castration and testosterone treatment on the vasoactive intestinal peptide receptor/effector system in rat prostatic epithelial cells. Biochim Biophys Acta 969:86–90

Carmena MJ, Sancho JI, Prieto JC (1986a) Additive effect of VIP or isoproterenol on forskolin-stimulated cyclic AMP accumulation in rat prostatic epithelial cells. Biochem Int 13:479–485

Carmena MJ, Sancho JI, Prieto JC (1986b) Effects of age and androgens upon functional vasoactive intestinal peptide receptors in rat prostatic epithelial cells. Biochim Biophys Acta 888:338–343

Casas AP (1958) The innervation of the human prostate. Z Mikrosk Anat Forsch 64:608–633

Chapple CR, Crowe R, Gilpin SA, Burnstock G (1991) The innervation of the human prostate gland—the changes associated with benign enlargement. J Urol 146:1637–1644

Chow PH, Dockery P, Cheung A (1997) Innervation of accessory sex glands in the adult male golden hamster and quantitative changes of nerve densities with age. Andrologia 29:331–342

Cockett AT, di Sant'Agnese PA, Gopinath P, Schoen SR, Abrahamsson PA (1993) Relationship of neuroendocrine cells of prostate and serotonin to benign prostatic hyperplasia. Urology 42:512–519

Cockle SM, Prater GV, Thetford CR, Hamilton C, Malone PR, Mundy AR (1994) Peptides related to thyrotrophin-releasing hormone (TRH) in human prostate and semen. Biochim Biophys Acta 1227:60–66

Cohen RJ, Glezerson G, Taylor LF, Grundle HA, Naude JH (1993) The neuroendocrine cell population of the human prostate gland. J Urol 150:365–368

Cowen T (1993) Ageing in the autonomic nervous system: a result of nerve-target interactions? A review. Mech Ageing Dev 68:163–173

Crowe R, Chapple CR, Burnstock G (1991) The human prostate gland: a histochemical and immunohistochemical study of neuropeptides, serotonin, dopamine beta-hydroxylase and acetylcholinesterase in autonomic nerves and ganglia. Br J Urol 68:53–61

Cunha GR, Donjacour AA, Cooke PS, Mee S, Bigsby RM, Higgins SJ, Sugimura Y (1987) The endocrinology and developmental biology of the prostate. Endocr Rev 8:338–362

Cunha GR, Donjacour AA, Sugimura Y (1986) Stromal-epithelial interactions and heterogeneity of proliferative activity within the prostate. Biochem Cell Biol 64:608–614

Dail WG (1993) Autonomic innervation of male reproductive genitalia. In: Maggi CA (ed.) Nervous control of the urogenital system. Harwood Academic Publisher. Switzerland. 69–101

Davis NS, Ewing JF, Mooney RA (1989) The neuroendocrine prostate: characterization and quantitation of calcitonin in the human gland. J Urol 142:884–888

DeKlerk DP, Coffey DS (1978) Quantitative determination of prostatic epithelial and stromal hyperplasia by a new technique. Biomorphometrics. Invest Urol 16:240–245

DeLellis RA (2001) The neuroendocrine system and its tumors: an overview. Am J Clin Pathol 115 Suppl: S5–S16

DeLellis RA, Dayal Y (1997) Neuroendocrine system. In: Sternberg SS (ed.) Histology for Pathologists. 2nd Edition. Lippincott Raven Publishers. Philadelphia. 1133–1151

DeLellis RA, Dayal Y, Tischler AS, Lee AK, Wolfe HJ (1986) Multiple endocrine neoplasia (MEN) syndromes: cellular origins and interrelationships. Int Rev Exp Pathol 28:163–215

DeLellis RA, Wolfe HJ (1981) The pathobiology of the human calcitonin (C)-cell: a review. Pathol Annu 16:25–52

di Sant'Agnese PA (1986) Calcitonin-like immunoreactive and bombesin-like immunoreactive endocrine-paracrine cells of the human prostate. Arch Pathol Lab Med 110:412–415

di Sant'Agnese PA (1992) Neuroendocrine differentiation in human prostatic carcinoma. Hum Pathol 23:287–296

di Sant'Agnese PA (1995) Neuroendocrine differentiation in prostatic carcinoma. Recent findings and new conepts. Cancer (Supplement) 75:1850–1859

di Sant'Agnese PA (1996) Neuroendocrine differentiation in the precursors of prostate cancer. Eur Urol 30:185–190

di Sant'Agnese PA (1998) Neuroendocrine cells of the prostate and neuroendocrine differentiation in prostatic carcinoma: a review of morphologic aspects. Urology 51:121–124

di Sant'Agnese PA (2001) Neuroendocrine differentiation in prostatic carcinoma: an update on recent developments. Ann Oncol 12 Suppl 2: S135–S140

di Sant'Agnese PA, Cockett AT (1994) The prostatic endocrine-paracrine (neuroendocrine) regulatory system and neuroendocrine differentiation in prostatic carcinoma: a review and future directions in basic research. J Urol 152:1927–1931

di Sant'Agnese PA, Davis NS, Chen M, Mesy Jensen KL (1987) Age-related changes in the neuroendocrine (endocrine-paracrine) cell population and the serotonin content of the guinea pig prostate. Lab Invest 57:729–736

di Sant'Agnese PA, Mesy Jensen KL (1984a) Endocrine-paracrine cells of the prostate and prostatic urethra: an ultrastructural study. Hum Pathol 15:1034–1041

di Sant'Agnese PA, Mesy Jensen KL (1984b) Somatostatin and/or somatostatin-like immunoreactive endocrine-paracrine cells in the human prostate gland. Arch Pathol Lab Med 108:693–696

di Sant'Agnese PA, Mesy Jensen KL (1987) Neuroendocrine differentiation in prostatic carcinoma. Hum Pathol 18:849–856

di Sant'Agnese PA, Mesy Jensen KL, Ackroyd RK (1989) Calcitonin, katacalcin, and calcitonin gene-related peptide in the human prostate. An immunocytochemical and immunoelectron microscopic study. Arch Pathol Lab Med 113:790–796 di Sant'Agnese PA, Mesy Jensen KL, Churukian CJ, Agarwal MM (1985) Human prostatic endocrine-paracrine (APUD) cells. Distributional analysis with a comparison of serotonin and neuron-specific enolase immunoreactivity and silver stains. Arch Pathol Lab Med 109:607–612

Di Silverio F, D'Eramo G, Flammia GP, Caponera M, Frascaro E, Buscarini M, Mariani M, Sciarra A (1993) Pathology of BPH. Minerva Urol Nefrol 45:135–142

Dinis P, Charrua A, Avelino A, Nagy I, Quintas J, Ribau U, Cruz F (2005) The distribution of sensory fibers immunoreactive for the TRPV1 (capsaicin) receptor in the human prostate. Eur Urol 48:162–167

Dizeyi N, Bjartell A, Nilsson E, Hansson J, Gadaleanu V, Cross N, Abrahamsson PA (2004) Expression of serotonin receptors and role of serotonin in human prostate cancer tissue and cell lines. Prostate 59:328–336

Donjacour AA, Cunha GR (1988) The effect of androgen deprivation on branching morphogenesis in the mouse prostate. Dev Biol 128:1-14

Edyvane KA, Smet PJ, Jonavicius J, Marshall VR (1995) Regional differences in the innervation of the human ureterovesical junction by tyrosine hydroxylase-, vasoactive intestinal peptide- and neuropeptide Y-like immunoreactive nerves. J Urol 154:262–268

Ekblad E, Edvinsson L, Wahlestedt C, Uddman R, Hakanson R, Sundler F (1984) Neuropeptide Y co-exists and co-operates with noradrenaline in perivascular nerve fibers. Regul Pept 8:225–235

Evangelou AI, Winter SF, Huss WJ, Bok RA, Greenberg NM (2004) Steroid hormones, polypeptide growth factors, hormone refractory prostate cancer, and the neuroendocrine phenotype. J Cell Biochem 91:671–683

Fahrenkrug J, Palle C, Jorgensen J, Ottesen B (1989) Regulatory peptides in the mammalian urogenital system. In: Polak JM (ed.) Regulatory peptides. Birkhäuser Verlag. Boston, Berlin. 362–381

Falck B, Owman CA (1965) A detailed methodological description of the fluorescence method for the cellular distribution of biogenic monoamines. Acta Univ Lund 7:5-23

Fawcett DW, Raviola E (1994) Bloom and Fawcett, a textbook of histology. Chapman and Hall. London, New York

Feldman SA, Eiden LE (2003) The chromogranins: their roles in secretion from neuroendocrine cells and as markers for neuroendocrine neoplasia. Endocr Pathol 14:3-23

Ferrari SL, Rizzoli R, Bonjour JP (1992) Parathyroid hormone-related protein production by primary cultures of mammary epithelial cells. J Cell Physiol 150:304–311

Fetissof F, Dubois MP, Arbeille-Brassart B, Lanson Y, Boivin F, Jobard P (1983) Endocrine cells in the prostate gland, urothelium and Brenner tumors. Immunohistological and ultrastructural studies. Virchows Arch B Cell Pathol Incl Mol Pathol 42:53–64

Feyrter F (1938) Uber diffuse endokrine epithaliale organe. Leipzig, Germany

Fisher DA, Lakshmanan J (1990) Metabolism and effects of epidermal growth factor and related growth factors in mammals. Endocr Rev 11:418–442

Fitzpatrick JM (2006) The natural history of benign prostatic hyperplasia. BJU Int 97 Suppl 2:3-6

Flickinger CJ (1972) The fine structure of the interstitial tissue of the rat prostate. Am J Anat 134:107–125

Frolich F (1949) Die "Helle Zelle" der bronchialschleinhaut und ihre beziehungen zum problem der chemoreceptoren. Frankfurter Z Pathol 60:517–558

Fujita T (1977) Concept of paraneurons. Arch Histol Jpn 40 Suppl: 1–12

Fujita T (1989) Present status of paraneuron concept. Arch Histol Cytol 52 Suppl: 1–8

Fujita T, Kanno T, Kobayashi S (1988) The Paraneuron. Springer Verlag. Tokyo

Fujita T, Kobayashi S, Yui R (1980) Paraneuron concept and its current implications. Adv Biochem Psychopharmacol 25:321–325

George FW, Peterson KG (1988) 5 alpha-dihydrotestosterone formation is necessary for embryogenesis of the rat prostate. Endocrinology 122:1159–1164

Giambanco I, Bianchi R, Ceccarelli P, Pula G, Sorci G, Antonioli S, Bocchini V, Donato R (1991) 'Neuron-specific' protein gene product 9.5 (PGP 9.5) is also expressed in glioma cell lines and its expression depends on cellular growth state. FEBS Lett 290:131–134

Gilpin SA, Gilpin CJ, Dixon JS, Gosling JA, Kirby RS (1986) The effect of age on the autonomic innervation of the urinary bladder. Br J Urol 58:378–381

Gkonos PJ, Krongrad A, Roos BA (1995a) Neuroendocrine peptides in the prostate. Urol Res 23:81–87

Gkonos PJ, Lokeshwar BL, Balkan W, Roos BA (1995b) Neuroendocrine peptides stimulate adenyl cyclase in normal and malignant prostate cells. Regul Pept 59:43–51

Golomb E, Rosenzweig N, Eilam R, Abramovici A (2000) Spontaneous hyperplasia of the ventral lobe of the prostate in aging genetically hypertensive rats. J Androl 21:58–64

Gould VE, Lee I, Wiedenmann B, Moll R, Chejfec G, Franke WW (1986) Synaptophysin: a novel marker for neurons, certain neuroendocrine cells, and their neoplasms. Hum Pathol 17:979–983

Gould VE, Moll R, Moll I, Lee I, Franke WW (1985) Neuroendocrine (Merkel) cells of the skin: hyperplasias, dysplasias, and neoplasms. Lab Invest 52:334–353

Greene EC (1955) Anatomy of the rat: Circulatory system. Hafner. New York. 200–203

Gu J, Polak JM, Probert L, Islam KN, Marangos PJ, Mina S, Adrian TE, McGregor GP, O'Shaughnessy DJ, Bloom SR (1983) Peptidergic innervation of the human male genital tract. J Urol 130:386–391

Gunn SA, Goud TC (1957) A correlative anatomical and functional study of the dorsolateral prostate of the rat. Anat Rec 128:41–53

Gutierrez-Canas I, Juarranz MG, Collado B, Rodriguez-Henche N, Chiloeches A, Prieto JC, Carmena MJ (2005) Vasoactive intestinal peptide induces neuroendocrine differentiation in the LNCaP prostate cancer cell line through PKA, ERK, and PI3K. Prostate 63:44–55

Guy L, Begin LR, Al Othman K, Chevalier S, Aprikian AG (1998) Neuroendocrine cells of the verumontanum: a comparative immunohistochemical study. Br J Urol 82:738–743

Hagn C, Schmid KW, Fischer-Colbrie R, Winkler H (1986) Chromogranin A, B, and C in human adrenal medulla and endocrine tissues. Lab Invest 55:405–411

Haimoto H, Takahashi Y, Koshikawa T, Nagura H, Kato K (1985) Immunohistochemical localization of gamma-enolase in normal human tissues other than nervous and neuroendocrine tissues. Lab Invest 52:257–263

Hamperl H (1932) Was sind argentaffine Zellen? Virchows Arch (A) 286:811–833

Harada Y, Okubo M, Yaga K, Kaneko T, Kaku K (1992) Neuropeptide Y inhibits beta-adrenergic agonist- and vasoactive intestinal peptide-induced cyclic AMP accumulation in rat pinealocytes through pertussis toxin-sensitive G protein. Neurochem 59:2178–2183

Hayashi N, Sugimura Y, Kawamura J, Donjacour AA, Cunha GR (1991) Morphological and functional heterogeneity in the rat prostatic gland. Biol Reprod 45:308–321

Heath DA, Senior PV, Varley JM, Beck F (1990) Parathyroid-hormone-related protein in tumours associated with hypercalcaemia. Lancet 335:66–69

Hedlund P, Ekstrom P, Larsson B, Alm P, Andersson KE (1997) Heme oxygenase and NO-synthase in the human prostate—relation to adrenergic, cholinergic and peptide-containing nerves. J Auton Nerv Syst 63:115–126

Heildenhain R (1870) Untersuchungen uber den Bau der Labdruesen. Arch Mikrosk Anat 6:368

Herrero M, Rodriguez A, Cejas H (2002) Prostatic neuroendocrine cells. Rev Fac Cien Med Univ Nac Cordoba 59:91–96

Higgins JR, Gosling JA (1989) Studies on the structure and intrinsic innervation of the normal human prostate. Prostate Suppl 2:5-16

Hokfelt T (1991) Neuropeptides in perspective: the last ten years. Neuron 7:867–879

Hokfelt T, Johansson O, Ljungdahl A, Lundberg JM, Schultzberg M (1980) Peptidergic neurones. Nature 284:515–521

Howard CV, Reed MG (2005a) Number estimation. In: Howard CV, Reed MG (eds.) Unbiased Stereology: three-dimensional measurement in microscopy. 2nd Edition. Bios Scientific Publishers. Oxford. 65–102

Howard CV, Reed MG (2005b) Length estimation. In: Howard CV, Reed MG (eds.) Unbiased Stereology: three-dimensional measurement in microscopy. 2nd Edition. Bios Scientific Publishers. Oxford. 119–125

Huggins C, Russel PS (1946) Quantitative effects of hypophysectomy on testis and prostate of dogs. Endocrinology 39:1-7

Hunter DJ, Berra-Unamuno A, Martin-Gordo A (1996) Prevalence of urinary symptoms and other urological conditions in Spanish men 50 years old or older. J Urol 155:1965–1970

Ingelmo I (2005) Efecto de la prolactina sobre la cantidad y distribución de las células neuroendocrinas y sobre la inervación peptidérgica en la próstata de la ratas normales y castradas. Tesis doctoral. Universidad Autónoma de Madrid. Facultad de Medina

Islam MA, Kato H, Hayama M, Kobayashi S, Igawa Y, Ota H, Nishizawa O (2002) Are neuroendocrine cells responsible for the development of benign prostatic hyperplasia? Eur Urol 42:79–83

Ito T, Yamamoto S, Ohno Y, Namiki K, Aizawa T, Akiyama A, Tachibana M (2001) Up-regulation of neuroendocrine differentiation in prostate cancer after androgen deprivation therapy: degree and androgen independence. Oncol Rep 8:1221–1224

Iwamura M, Abrahamsson PA, Benning CM, Cockett AT, di Sant'Agnese PA (1994a) Androgen receptor immunostaining and its tissue distribution in formalin-fixed, paraffin-embedded sections after microwave treatment. J Histochem Cytochem 42:783–788

Iwamura M, di Sant'Agnese PA, Wu G, Benning CM, Cockett AT, Gershagen S (1994b) Overexpression of human epidermal growth factor receptor and c-erbB-2 by neuroendocrine cells in normal prostatic tissue. Urology 43:838–843

Iwamura M, Abrahamsson PA, Foss KA, Wu G, Cockett AT, Deftos LJ (1994c) Parathyroid hormone-related protein: a potential autocrine growth regulator in human prostate cancer cell lines. Urology 43:675–679

Iwamura M, Wu G, Abrahamsson PA, di Sant'Agnese PA, Cockett AT, Deftos LJ (1994d) Parathyroid hormone-related protein is expressed by prostatic neuroendocrine cells. Urology 43:667–674

Iwamura M, di Sant'Agnese PA, Wu G, Benning CM, Cockett AT, Deftos LJ, Abrahamsson PA (1993) Immunohistochemical localization of parathyroid hormone-related protein in human prostate cancer. Cancer Res 53:1724–1726

Iwasa A (1993) Distribution of neuropeptide Y (NPY) and its binding sites in human lower urinary tract. Histological analysis. Nippon Hinyokika Gakkai Zasshi 84:1000–1006

Iwata T, Ukimura O, Inaba M, Kojima M, Kumamoto K, Ozawa H, Kawata M, Miki T (2001) Immunohistochemical studies on the distribution of nerve fibers in the human prostate with special reference to the anterior fibromuscular stroma. Prostate 48:242–247

Jen PY, Dixon JS (1995) Development of peptide-containing nerves in the human fetal prostate gland. J Anat 187 (Pt 1): 169–179

Jesik CJ, Holland JM, Lee C (1982) An anatomic and histologic study of the rat prostate. Prostate 3:81–97

Jimenez N, Calvo A, Martinez A, Rosell D, Cuttitta F, Montuenga LM (1999) Expression of adrenomedullin and proadrenomedullin N-terminal 20 peptide in human and rat prostate. J Histochem Cytochem 47:1167–1178

Jongsma J, Oomen MH, Noordzij MA, Romijn JC, Der Kwast TH, Schroder FH, van Steenbrugge GJ (2000a) Androgen-independent growth is induced by neuropeptides in human prostate cancer cell lines. Prostate 42:34–44

Jongsma J, Oomen MH, Noordzij MA, Van Weerden WM, Martens GJ, van der Kwast TH, Schroder FH, van Steenbrugge GJ (2000b) Androgen deprivation of the PC-310 [correction of prohormone convertase-310] human prostate cancer model system induces neuroendocrine differentiation. Cancer Res 60:741–748

Juarranz MG, Bodega G, Prieto JC, Guijarro LG (2001) Vasoactive intestinal peptide (VIP) stimulates rat prostatic epithelial cell proliferation. Prostate 47:285–292

Juarranz MG, Guijarro LG, Bajo AM, Carmena MJ, Prieto JC (1994) Ontogeny of vasoactive intestinal peptide receptors in rat ventral prostate. Gen Pharmacol 25:509–514

Kawano H, Daikoku S, Saito S (1983) Location of thyrotropin-releasing hormone-like immunoreactivity in rat pancreas. Endocrinology 112:951–955

Keast JR (1995) Visualization and immunohistochemical characterization of sympathetic and parasympathetic neurons in the male rat major pelvic ganglion. Neuroscience 66:655–662

Kepper ME, Keast JR (1995) Immunohistochemical properties and spinal connections of pelvic autonomic neurons that innervate the rat prostate gland. Cell Tissue Res 281:533–542

Kerr JF, Searle J (1973) Deletion of cells by apoptosis during castration-induced involution of the rat prostate. Virchows Arch B Cell Pathol 13:87–102

Killam AL, Watts SW, Cohen ML (1995) Role of α_1-adrenoceptors and 5-HT$_2$ receptors in serotonin-induced contraction of rat prostate: autoradiographical and functional studies. Eur J Pharmacol 273:7–14

Kinbara H, Cunha GR (1996) Ductal heterogeneity in rat dorsal-lateral prostate. Prostate 28:58–64

Kirby RS (1992) The clinical assessment of benign prostatic hyperplasia. Cancer 70:284–290

Kollermann J, Helpap B (2001) Neuroendocrine differentiation and short-term neoadjuvant hormonal treatment of prostatic carcinoma with special regard to tumor regression. Eur Urol 40:313–317

Krijnen JL, Janssen PJ, Ruizeveld de Winter JA, van Krimpen H, Schroder FH, van der Kwast TH (1993) Do neuroendocrine cells in human prostate cancer express androgen receptor? Histochemistry 100:393–398

Kurimoto S, Moriyama N, Hamada K, Kawabe K (1998) Evaluation of histological structure and its effect on the distribution of α_1-adrenoceptors in human benign prostatic hyperplasia. Br J Urol 81:388–393

Laczko I, Hudson DL, Freeman A, Feneley MR, Masters JR (2005) Comparison of the zones of the human prostate with the seminal vesicle: morphology, immunohistochemistry, and cell kinetics. Prostate 62:260–266

Lange W, Unger J (1990) Peptidergic innervation within the prostate gland and seminal vesicle. Urol Res 18:337–340

Larsen JJ, Ottesen B, Fahrenkrug J (1981) Vasoactive intestinal polypeptide (VIP) in the male genitourinary tract: concentration and motor effect. Invest Urol 19:211–213

Lau WA, Pennefather JN (1998) Muscarinic receptor subtypes in the rat prostate gland. Eur J Pharmacol 343:151–156

Lau WA, Ventura S, Pennefather JN (1998) Pharmacology of neurotransmission to the smooth muscle of the rat and the guinea-pig prostate glands. J Auton Pharmacol 18:349–356

Lauweryns JM, Peuskens JC (1972) Neuro-epithelial bodies (neuroreceptor or secretory organs?) in human infant bronchial and bronchiolar epithelium. Anat Rec 172:471–481

Le Douarin NM, Creuzet S, Couly G, Dupin E (2004) Neural crest cell plasticity and its limits. Development 131:4637–4650

Le Douarin NM, Dupin E (2003) Multipotentiality of the neural crest. Curr Opin Genet Dev 13:529–536

Le Douarin NM, Kalcheim C (1999) The neural crest. Cambridge University Press. New York

Lee C (1996) Role of androgen in prostate growth and regression: stromal-epithelial interaction. Prostate Suppl 6:52–56

Lee C, Goolsby CL, Sensibar JA (1994) Cell cycle kinetics in rat prostatic epithelia: nuclear migration during G_2 phase. J Urol 152:2294–2299

Lee C, Kozlowski JM, Grayhack JT (1997) Intrinsic and extrinsic factors controlling benign prostatic growth. Prostate 31:131–138

Lee C, Sensibar JA, Dudek SM, Hiipakka RA, Liao ST (1990) Prostatic ductal system in rats: regional variation in morphological and functional activities. Biol Reprod 43:1079–1086

Leissner KH, Tisell LE (1979) The weight of the human prostate. Scand J Urol Nephrol 13:137–142

Luca IC (1998) Endocrine diffuse system. Histological and functional aspects interrelated with tumoral pathology. Rom J Morphol Embryol 44:17–22

Maini A, Archer C, Wang CY, Haas GP (1997) Comparative pathology of benign prostatic hyperplasia and prostate cancer. In Vivo 11:293–299

Mao P, Angrist A (1966) The fine structure of the basal cell of human prostate. Lab Invest 15:1768–1782

Marraco G (2004) Neuroendocrine differentiation in prostate carcinoma. Review. Rev Arg Urol 69:1-5

Martin R, Fraile B, Peinado F, Arenas MI, Elices M, Alonso L, Paniagua R, Martin JJ, Santamaria L (2000) Immunohistochemical localization of protein gene product 9.5, ubiquitin, and neuropeptide Y immunoreactivities in epithelial and neuroendocrine cells from normal and hyperplastic human prostate. J Histochem Cytochem 48:1121–1130

Martin R, Santamaria L, Fraile B, Paniagua R, Polak JM (1995) Ultrastructural localization of PGP 9.5 and ubiquitin immunoreactivities in rat ductus epididymidis epithelium. Histochem J 27:431–439

Martinez-Pineiro L, Dahiya R, Nunes LL, Tanagho EA, Schmidt RA (1993) Pelvic plexus denervation in rats causes morphologic and functional changes of the prostate. J Urol 150:215–218

Masson P (1914) Le gland endocrine de l'intestine chez l'homme. C R Acad Sci Paris 59

Mayhew TM, Gundersen HJ (1996) If you assume, you can make an ass out of u and me: a decade of the disector for stereological counting of particles in 3D space. J Anat 188 (Pt 1): 1–15

McCarty KSJr, Nath M (1997) Breast. In: Sternberg SS (ed.) Histology for pathologists. 2nd Edition. Lippincott Raven Publishers. Philadelphia. 71–82

McNeal JE (1981) The zonal anatomy of the prostate. Prostate 2:35–49

McNeal JE (1984) Anatomy of the prostate and morphogenesis of BPH. Prog Clin Biol Res 145:27–53

McNeal JE (1990) Pathology of benign prostatic hyperplasia. Insight into etiology. Urol Clin North Am 17:477–486

McNeal JE (1997) Prostate. In:Histology for pathologists. 2nd Edition. Lippincott Raven Publishers. Philadelphia. 997–1017

McVary KT, McKenna KE, Lee C (1998) Prostate innervation. Prostate Suppl 8:2-13

McVary KT, Razzaq A, Lee C, Venegas MF, Rademaker A, McKenna KE (1994) Growth of the rat prostate gland is facilitated by the autonomic nervous system. Biol Reprod 51:99–107

Meyer JS, Sufrin G, Martin SA (1982) Proliferative activity of benign human prostate, prostatic adenocarcinoma and seminal vesicle evaluated by thymidine labeling. J Urol 128:1353–1356

Minker E, Bartha C (1981) Pharmacological responses by different portions of guinea pig vas deferens circular muscle preparation. Acta Physiol Acad Sci Hung 58:65–77

Moll I, Roessler M, Brandner JM, Eispert AC, Houdek P, Moll R (2005) Human Merkel cells—aspects of cell biology, distribution and functions. Eur J Cell Biol 84:259–271

Montuenga LM, Guembe L, Burrell MA, Bodegas ME, Calvo A, Sola JJ, Sesma P, Villaro AC (2003) The diffuse endocrine system: from embryogenesis to carcinogenesis. Prog Histochem Cytochem 38:155–272

Mosca A, Dogliotti L, Berruti A, Lamberts SW, Hofland LJ (2004) Somatostatin receptors: from basic science to clinical approach. Unlabeled somatostatin analogues-1: Prostate cancer. Dig Liver Dis 36 Suppl 1: S60-S67

Mottet N, Costa P, Bali JP (1999) Autonomic nervous system and prostatic physiology. Specific features of the alpha-adrenergic system. Prog Urol 9:26–36

Nadelhaft I (2003) Cholinergic axons in the rat prostate and neurons in the pelvic ganglion. Brain Res 989:52–57

Nadelhaft I, Miranda-Sousa AJ, Vera PL (2002) Separate urinary bladder and prostate neurons in the central nervous system of the rat: simultaneous labeling with two immunohistochemically distinguishable pseudorabies viruses. BMC Neurosci 3:8-19

Nakada SY, di Sant'Agnese PA, Moynes RA, Hiipakka RA, Liao S, Cockett AT, Abrahamsson PA (1993) The androgen receptor status of neuroendocrine cells in human benign and malignant prostatic tissue. Cancer Res 53:1967–1970

Narayan P (1992) Neoplasias de la próstata. In: McAninch JW (ed.) Smith's general urology. Appleton and Lange

Nemeth JA, Lee C (1996) Prostatic ductal system in rats: regional variation in stromal organization. Prostate 28:124–128

Nevalainen MT, Valve EM, Ingleton PM, Harkonen PL (1996) Expression and hormone regulation of prolactin receptors in rat dorsal and lateral prostate. Endocrinology 137:3078–3088

Noordzij MA, van Steenbrugge GJ, van der Kwast TH, Schroder FH (1995) Neuroendocrine cells in the normal, hyperplastic and neoplastic prostate. Urol Res 22:333–341

Oesterling JE (1996) Benign prostatic hyperplasia: a review of its histogenesis and natural history. Prostate Suppl 6:67–73

Orr R, Marson L (1998) Identification of CNS neurons innervating the rat prostate: a transneuronal tracing study using pseudorabies virus. J Auton Nerv Syst 72:4-15

Partin AW, Oesterling JE, Epstein JI, Horton R, Walsh PC (1991) Influence of age and endocrine factors on the volume of benign prostatic hyperplasia. J Urol 145:405–409

Pearse AG (1969) The cytochemistry and ultrastructure of polypeptide hormone-producing cells of the APUD series and the embryologic, physiologic and pathologic implications of the concept. J Histochem Cytochem 17:303–313

Pearse AG, Polak JM (1971) Neural crest origin of the endocrine polypeptide (APUD) cells of the gastrointestinal tract and pancreas. Gut 12:783–788

Peehl DM (1996) Cellular biology of prostatic growth factors. Prostate Suppl 6:74–78

Peinado-Ibarra F (1998) La hiperplasia benigna de próstata: estudio estereológico e inmunohistoquímico de la proliferación y el número de células, el volumen de los compartimentos prostáticos y la expresión de TGFB1. Tesis doctoral. Universidad Autónoma de Madrid. Facultad de Medicina. Madrid

Pekary AE, Sharp B, Briggs J, Carlson HE, Hershman JM (1983) High concentrations of p-Glu-His-Pro-NH$_2$ (thyrotropin-releasing hormone) occur in rat prostate. Peptides 4:915–919

Pennefather JN, Lau WA, Mitchelson F, Ventura S (2000) The autonomic and sensory innervation of the smooth muscle of the prostate gland: a review of pharmacological and histological studies. J Auton Pharmacol 20:193–206

Perez Casas A, CaboTamargo J, BengoecheaGonzalez E, Lopez Muniz A, Vega Alvarez JA (1985) Microscopic innervation of the prostate. II. The intramural microganglia. Arch Esp Urol 38:365–373

Polak JM (1989) Regulatory peptides. Birkhäuser Verlag. Basel, Boston, Berlin

Polak JM, Bloom SR (1983) Regulatory peptides: key factors in the control of bodily functions. Br Med J (Clin Res Ed) 286:1461–1466

Polak JM, Van Noorden S (1988) Inmunoenzyme methods. In:An introduction to inmunochemistry: current techniques and problems. Oxford University Press. 19–23

Portela-Gomes GM, Stridsberg M, Johansson H, Grimelius L (1999) Co-localization of synaptophysin with different neuroendocrine hormones in the human gastrointestinal tract. Histochem Cell Biol 111:49–54

Powell MS, Li R, Dai H, Sayeeduddin M, Wheeler TM, Ayala GE (2005) Neuroanatomy of the normal prostate. Prostate 65:52–57

Pretl K (1944) Zur frage der endokrinei der menschlichen vorsteherdruse. Virchows Arch (A) 32:392–404

Prieto JC, Carmena MJ (1983) Receptors for vasoactive intestinal peptide on isolated epithelial cells of rat ventral prostate. Biochim Biophys Acta 763:408–413

Properzi G, Cordeschi G, Francavilla S (1992) Postnatal development and distribution of peptide-containing nerves in the genital system of the male rat. An immunohistochemical study. Histochemistry 97:61–68

Purinton PT, Fletcher TF, Bradley WE (1973) Gross and light microscopic features of the pelvic plexus in the rat. Anat Rec 175:697–705

Reese JH, McNeal JE, Redwine EA, Samloff IM, Stamey TA (1986) Differential distribution of pepsinogen II between the zones of the human prostate and the seminal vesicle. J Urol 136:1148–1152

Reese JH, McNeal JE, Redwine EA, Stamey TA, Freiha FS (1988) Tissue type plasminogen activator as a marker for functional zones, within the human prostate gland. Prostate 12:47–53

Reiter E, Kalhs P, Keil F, Rabitsch W, Gisslinger H, Mayer G, Worel N, Lechner K, Greinix HT (1999) Effect of high-dose melphalan and peripheral blood stem cell transplantation on renal function in patients with multiple myeloma and renal insufficiency: a case report and review of the literature. Ann Hematol 78:189–191

Rode J, Dhillon AP, Doran JF, Jackson P, Thompson RJ (1985) PGP 9.5, a new marker for human neuroendocrine tumours. Histopathology 9:147–158

Rodríguez Ramos, M. R (2001) Estudio inmunoshistoquímico y cuantitativo de las células neuroendocrinas y de la inervación peptidérgica en la próstata de la rata durante el desarrollo postnatal. Tesis doctoral. Universidad San Pablo CEU. Facultad de Ciencias Experimentales y de la Salud

Rodriguez R, Pozuelo JM, Martin R, Arriazu R, Santamaria L (2005) Stereological quantification of nerve fibers immunoreactive to PGP 9.5, NPY, and VIP in rat prostate during postnatal development. J Androl 26:197–204

Rodriguez R, Pozuelo JM, Martin R, Henriques-Gil N, Haro M, Arriazu R, Santamaria L (2003) Presence of neuroendocrine cells during postnatal development in rat prostate: immunohistochemical, molecular, and quantitative study. Prostate 57:176–185

Rumpold H, Heinrich E, Untergasser G, Hermann M, Pfister G, Plas E, Berger P (2002a) Neuroendocrine differentiation of human prostatic primary epithelial cells in vitro. Prostate 53:101–108

Rumpold H, Untergasser G, Madersbacher S, Berger P (2002b) The development of benign prostatic hyperplasia by trans-differentiation of prostatic stromal cells. Exp Gerontol 37:1001–1004

Sakamoto N, Hasegawa Y, Koga H, Kotoh S, Kuroiwa K, Naito S (1999) Presence of ganglia within the prostatic capsule: ganglion involvement in prostatic cancer. Prostate 40:167–171

Santamaria L, Martin R, Martin JJ, Alonso L (2002) Stereologic estimation of the number of neuroendocrine cells in normal human prostate detected by immunohistochemistry. Appl Immunohistochem Mol Morphol 10:275–281

Santamaria L, Martin R, Paniagua R, Fraile B, Nistal M, Terenghi G, Polak JM (1993) Protein gene product 9.5 and ubiquitin immunoreactivities in rat epididymis epithelium. Histochemistry 100:131–138

Schalken JA, van Leenders G (2003) Cellular and molecular biology of the prostate: stem cell biology. Urology 62:11–20

Scharrer B (1967) The neurosecretory neuron in neuroendocrine regulatory mechanisms. Am Zool 7:161–169

Schmid KW, Helpap B, Totsch M, Kirchmair R, Dockhorn-Dworniczak B, Bocker W, Fischer-Colbrie R (1994) Immunohistochemical localization of chromogranins A and B and secretogranin II in normal, hyperplastic and neoplastic prostate. Histopathology 24:233–239

Sciarra A, Cardi A, Dattilo C, Mariotti G, Di Monaco F, Di Silverio F (2006) New perspective in the management of neuroendocrine differentiation in prostate adenocarcinoma. Int J Clin Pract 60:462–470

Scolnik M, Tykochinsky G, Servadio C, Abramovici A (1992) The development of vascular supply of normal rat prostate during the sexual maturation: an angiographic study. Prostate 21:1-14

Sensibar JA, Griswold MD, Sylvester SR, Buttyan R, Bardin CW, Cheng CY, Dudek S, Lee C (1991) Prostatic ductal system in rats: regional variation in localization of an androgen-repressed gene product, sulfated glycoprotein-2. Endocrinology 128:2091–2102

Serezhin BS (1988) APUD cells in basal cell proliferative tissue of the prostate. Arkh Patol 50:46–51

Shabsigh A, Tanji N, D'Agati V, Burchardt T, Burchardt M, Hayek O, Shabsigh R, Buttyan R (1999) Vascular anatomy of the rat ventral prostate. Anat Rec 256:403–411

Sherwood ER, Lee C (1995) Epidermal growth factor-related peptides and the epidermal growth factor receptor in normal and malignant prostate. World J Urol 13:290–296

Sidhu GS, Chandra P, Cassai ND (2005) Merkel cells, normal and neoplastic: an update. Ultrastruct Pathol 29:287–294

Slovin SF (2006) Neuroendocrine differentiation in prostate cancer: a sheep in wolf's clothing? Nat Clin Pract Urol 3:138–144

Smet PJ, Edyvane KA, Jonavicius J, Marshall VR (1994) Colocalization of nitric oxide synthase with vasoactive intestinal peptide, neuropeptide Y, and tyrosine hydroxylase in nerves supplying the human ureter. J Urol 152:1292–1296

Snyder SH (1980) Brain peptides as neurotransmitters. Science 209:976–983

Solano RM, Carmena MJ, Carrero I, Cavallaro S, Roman F, Hueso C, Travali S, Lopez-Fraile N, Guijarro LG, Prieto JC (1996) Characterization of vasoactive intestinal peptide/pituitary adenylate cyclase-activating peptide receptors in human benign hyperplastic prostate. Endocrinology 137:2815–2822

Solano RM, Carmena MJ, Guijarro LG, Prieto JC (1994) Neuropeptide Y inhibits vasoactive intestinal peptide-stimulated adenylyl cyclase in rat ventral prostate. Neuropeptides 27:31–37

Sommerfeld HJ, Partin AW, Pannek J (2002) Incidence of neuroendocrine cells in the seminal vesicles and the prostate—an immunohistochemical study. Int Urol Nephrol 34:357–360

Sugimura Y, Cunha GR, Donjacour AA (1986) Morphogenesis of ductal networks in the mouse prostate. Biol Reprod 34:961–971

Tainio H (1995) Peptidergic innervation of the human prostate, seminal vesicle and vas deferens. Acta Histochem 97:113–119

Tanagho EA (1982) Embriologic development of the urinary tract. In: AUA Update series. American Urological Association. 1–8

Theodoropoulos VE, Tsigka A, Mihalopoulou A, Tsoukala V, Lazaris AC, Patsouris E, Ghikonti I (2005) Evaluation of neuroendocrine staining and androgen receptor expression in incidental prostatic adenocarcinoma: prognostic implications. Urology 66:897–902

Tilley WD, Buchanan G, Hickey TE, Bentel JM (1996) Mutations in the androgen receptor gene are associated with progression of human prostate cancer to androgen independence. Clin Cancer Res 2:277–285

Timms BG, Mohs TJ, Didio LJ (1994) Ductal budding and branching patterns in the developing prostate. J Urol 151:1427–1432

Toni R (2004) The neuroendocrine system: organization and homeostatic role. J Endocrinol Invest 27:35–47

Torres G, Bitran M, Huidobro-Toro JP (1992) Co-release of neuropeptide Y (NPY) and noradrenaline from the sympathetic nerve terminals supplying the rat vas deferens; influence of calcium and the stimulation intensity. Neurosci Lett 148:39–42

Uchida K, Masumori N, Takahashi A, Itoh N, Tsukamoto T (2005) Characterization of prostatic neuroendocrine cell line established from neuroendocrine carcinoma of transgenic mouse allograft model. Prostate 62:40–48

Untergasser G, Madersbacher S, Berger P (2005) Benign prostatic hyperplasia: age-related tissue-remodeling. Exp Gerontol 40:121–128

Vaalasti A, Hervonen A (1979) Innervation of the ventral prostate of the rat. Am J Anat 154:231–243

Vaalasti A, Linnoila I, Hervonen A (1980) Immunohistochemical demonstration of VIP, [Met5]- and [Leu5]-enkephalin immunoreactive nerve fibres in the human prostate and seminal vesicles. Histochemistry 66:89–98

Van Lommel A, Bolle T, Fannes W, Lauweryns JM (1999) The pulmonary neuroendocrine system: the past decade. Arch Histol Cytol 62:1-16

Vega JA, Zubizarreta JF, del Valle M, Hernandez LC, Perez-Casas A (1990) Vasoactive intestinal polypeptide (VIP)-like immunoreactivity in intraprostatic neurons and major pelvic ganglia in the rat. Arch Esp Urol 43:93–96

Ventura S, Pennefather J, Mitchelson F (2002) Cholinergic innervation and function in the prostate gland. Pharmacol Ther 94:93–112

Vittoria A, La Mura E, Cocca T, Cecio A (1990) Serotonin-, somatostatin- and chromogranin A-containing cells of the urethro-prostatic complex in the sheep. An immunocytochemical and immunofluorescent study. J Anat 171:169–178

Walsh PC, Retik AB, Vaughan ED, Wein AJ, Kavoussi LR, Novick AC, Partin AW, Peters CA (2002) Campbell's urology. W. B. Saunders. Philadelphia

Wang JM, McKenna KE, Lee C (1991a) Determination of prostatic secretion in rats: effect of neurotransmitters and testosterone. Prostate 18:289–301

Wang JM, McKenna KE, McVary KT, Lee C (1991b) Requirement of innervation for maintenance of structural and functional integrity in the rat prostate. Biol Reprod 44:1171–1176

Wanigasekara Y, Kepper ME, Keast JR (2003) Immunohistochemical characterisation of pelvic autonomic ganglia in male mice. Cell Tissue Res 311:175–185

Warburton AL, Santer RM (1994) Sympathetic and sensory innervation of the urinary tract in young adult and aged rats: a semi-quantitative histochemical and immunohistochemical study. Histochem J 26:127–133

Wilkinson KD, Lee KM, Deshpande S, Duerksen-Hughes P, Boss JM, Pohl J (1989) The neuron-specific protein PGP 9.5 is a ubiquitin carboxyl-terminal hydrolase. Science 246:670–673

Wu G, Burzon DT, di Sant'Agnese PA, Schoen S, Deftos LJ, Gershagen S, Cockett AT (1996) Calcitonin receptor mRNA expression in the human prostate. Urology 47:376–381

Xing N, Qian J, Bostwick D, Bergstralh E, Young CY (2001) Neuroendocrine cells in human prostate over-express the anti-apoptosis protein survivin. Prostate 48:7–15

Xue Y, Sonke G, Schoots C, Schalken J, Verhofstad A, de la RJ, Smedts F (2001) Proliferative activity and branching morphogenesis in the human prostate: a closer look at pre- and postnatal prostate growth. Prostate 49:132–139

Xue Y, van der LJ, Smedts F, Schoots C, Verhofstad A, de la RJ, Schalken J (2000) Neuroendocrine cells during human prostate development: does neuroendocrine cell density remain constant during fetal as well as postnatal life? Prostate 42:116–123

Xue Y, Verhofstad A, Lange W, Smedts F, Debruyne F, de la RJ, Schalken J (1997) Prostatic neuroendocrine cells have a unique keratin expression pattern and do not express Bcl-2: cell kinetic features of neuroendocrine cells in the human prostate. Am J Pathol 151:1759–1765

Yokoyama R, Inokuchi T, Satoh H, Kusaba T, Yamamoto K, Ando K (1990) Distribution of tyrosine hydroxylase (TH)-like, neuropeptide Y (NPY)-like immunoreactive and acetylcholinesterase (AChE)-positive nerve fibers in the prostate gland of the monkey (*Macacus fuscatus*) Kurume Med J 37:1–8

Young B, Lowe JS, Stevens A, Heath JW, Wheater PR (2006) Wheater's functional histology, a text and colour atlas. Hardcourt, Churchill Livingstone. London

Zermann DH, Ishigooka M, Doggweiler R, Schubert J, Schmidt RA (2000) Central nervous system neurons labeled following the injection of pseudorabies virus into the rat prostate gland. Prostate 44:240–247

Zhu J, Li W, Toews ML, Hexum TD (1992) Neuropeptide Y inhibits forskolin-stimulated adenylate cyclase in bovine adrenal chromaffin cells via a pertussis toxin-sensitive process. J Pharmacol Exp Ther 263:1479–1486

Subject Index